T5-CUL-548

A C S S Y M P O S I U M S E R I E S **239**

Assessment and Management of Chemical Risks

Joseph V. Rodricks, EDITOR
Environ Corporation

Robert G. Tardiff, EDITOR
Life Systems, Inc.

Based on a symposium sponsored by
the Division of Chemical Health and Safety
at the 184th Meeting
of the American Chemical Society,
Kansas City, Missouri,
September 12–17, 1982

American Chemical Society, Washington, D.C. 1984

Library of Congress Cataloging in Publication Data

Assessment and management of chemical risks.
(ACS symposium series, ISSN 0097-6156; 239)

Bibliography: p.
Includes indexes.

Contents: Conceptual basis for risk assessment/ Joseph V. Rodricks and Robert G. Tardiff—Use of toxicity test data in the estimation of risks to human health/Norton Nelson—Interspecies extrapolation/ Daniel B. Menzel and Elaine D. Smolko—[etc.]

1. Toxicity testing—Congresses. 2. Toxicology—Congresses.

I. Rodricks, Joseph V., 1938- . II. Tardiff, Robert G. III. American Chemical Society. Division of Chemical Health and Safety. IV. American Chemical Society. Meeting (184th: 1982: Kansas City, Mo.) V. Series.

RA1199.A77 1984 363.1'79 83-25851
ISBN 0-8412-0821-2

PRINTED IN THE UNITED STATES OF AMERICA

Second printing 1985

ACS Symposium Series

M. Joan Comstock, *Series Editor*

FOREWORD

The ACS SYMPOSIUM SERIES was founded in 1974 to provide a medium for publishing symposia quickly in book form. The format of the Series parallels that of the continuing ADVANCES IN CHEMISTRY SERIES except that in order to save time the papers are not typeset but are reproduced as they are submitted by the authors in camera-ready form. Papers are reviewed under the supervision of the Editors with the assistance of the Series Advisory Board and are selected to maintain the integrity of the symposia; however, verbatim reproductions of previously published papers are not accepted. Both reviews and reports of research are acceptable since symposia may embrace both types of presentation.

CONTENTS

PREFACE

THE CHIEF GOAL OF CHEMICAL RISK ASSESSMENT is to characterize the types of hazards associated with a substance and to estimate the probability that those hazards will be realized in exposed populations or individuals. Risk assessment is distinct from risk management, which is the process of deciding how best to mitigate risks deemed to be excessive. Risk assessment depends upon data derived from experimental and epidemiological investigations into the hazardous properties of chemicals and from studies of the magnitude of human exposure to them. Risk management decisions are influenced by judgments about the importance of an assessed risk to public health, the technical means by which a risk might be abated and the costs of such abatement, and the applicable laws. Risk management decisions may thus take a wide variety of forms and depend upon many factors that exceed the bounds of science.

The chapters in this book concern both risk assessment and risk management. The first five deal with some of the central problems of risk assessment. The remaining six chapters cover a range of risk management topics, and reveal some of the principal issues facing chemical risk managers in a number of different contexts. The chapters on risk management discuss the pervasive problem of dealing with the scientific uncertainties associated with assessed risks, the use of comparative risk analysis as a basis for deciding whether risk controls should be sought, and the legal issues that always need to be considered. The chapters on risk management also reveal some of the fundamental problems faced by both corporate and regulatory decision makers.

This book is by no means a comprehensive treatise on either the assessment or management of chemical risks. Rather, it is an introduction to the essential elements of these subjects, designed especially for the increasing number of individuals, particularly those in the corporate setting, who are having to make decisions about chemical risks in the face of substantial scientific uncertainty and without the benefit of strong historical precedents. We hope this volume serves to lay the groundwork for an understanding of these issues and to stimulate further inquiry.

JOSEPH V. RODRICKS
Environ Corporation
Washington, DC

ROBERT G. TARDIFF
Life Systems, Inc.
Arlington, VA

October 12, 1983

ASSESSMENT OF CHEMICAL RISKS

1

Conceptual Basis for Risk Assessment

JOSEPH V. RODRICKS

Environ Corporation, Washington, DC 20006

ROBERT G. TARDIFF

Board on Toxicology and Environmental Health Hazards, National Academy of Sciences/ National Research Council, Washington, DC 20037

Risk is the probability of injury or death. For some activities we encounter no great difficulties in determining risk. Thus, it is possible to estimate quite accurately the risks of accidental death due to such activities as driving a car, working in a coal mine, riding a bicycle, hiking in the desert, or eating low-acid canned foods (botulism). Estimation of such risks is readily accomplished because historical statistical data are available, and because there is little difficulty in demonstrating the causal connections between injury and these types of activities. To estimate such risks is the work of actuaries, most of whom are employed by insurance companies.

Other risks cannot be so easily estimated because the necessary actuarial data do not exist and frequently cannot even be collected. Many of the potential risks from exposure to chemicals are in this second category. In addition to the absence of actuarial data relating to them, these risks tend to have the following characteristics:

(i) Suspicion that exposure may lead to injury usually results from experimental observations, commonly involving animals.

(ii) Identifiable injury does not occur immediately following exposure, and may sometimes not occur for many years after initial exposure.

0097-6156/84/0239-0003$06.00/0

(iii) The conditions of exposure (level, frequency, duration, route) that give rise to experimentally-observed injury are frequently different (sometimes radically so) from the conditions of actual human exposure, which themselves may not be well-defined.

(iv) The experimental environments in which information is collected on potential injury from a chemical exposure are usually free of the large number of factors in the human environment that may biologically or chemically interact with the chemical, and thus alter its capacity to cause injury.

(v) Experiments used to collect data on chemical injury may involve several different species of test animals, and they may yield quantitatively, and sometimes qualitatively, different results. It is usually not feasible to identify the species that best mimics human response, assuming there is one at all.

(vi) Epidemiological investigations of chronic exposure or injury, while yielding data on the species of concern, are frequently limited because they can not usually detect small but possibly important effects; because they frequently can not provide evidence of strict causation; and because they usually do not provide quantitative dose-response data. Moreover, they can be conducted only after exposure has occurred and thus can not be used to decide whether exposure to a newly-introduced substance should be permitted.

Given the above, it would seem foolish to attempt to predict the human risks associated with exposures to chemicals. Many scientists faced with such a problem are not willing to attempt an answer, and proclaim the need for more research. They believe that it would be scientifically imprudent ever to go beyond the empirical data to predict risks under different conditions. This belief ignores the possibility that low but nonetheless important risks exist under conditions of exposure that defy our attempts at direct observation. In addition, in the context of current law, such a view automatically translates to a regulatory decision to permit exposures to continue or to begin, because there would be no reason to limit exposures at any level below those for which empirical information on health effects is available. In light of current knowledge this could be a highly imprudent public health policy.

If we fail to find workable approaches to the problem of assessing chemical risk, and fail to identify some systematic way to deal with these scientific uncertainties, we would indeed find ourselves in a serious predicament. Thus, we would be faced with the prospect of not being able to decide whether exposure to a chemical can or can not be permitted, unless we base the decision on grounds completely unrelated to the question of risk. The latter course seems highly undesirable, although it has sometimes been taken.[1]

In the context of regulatory decision-making, the difficulties of defining the nature and magnitude of chemical risk can be overcome (indeed, have been for years) by the application of certain operational schemes. Application of these schemes can not be claimed to lead to true estimates of human risk, yet there are good reasons to believe that they meet the desirable criterion of being capable of distinguishing low from high risk exposures, and do so in a systematic fashion.

The major operational schemes now in use represent two strikingly different approaches to the problem of assessing the health consequences of chemical exposures, and we shall now describe them.

Traditional Safety Assessment Schemes

The task of assigning safe exposure levels for chemicals has traditionally been assigned to toxicologists. During the first half of this century, this problem arose in connection with food additives, pesticides, drugs, and occupational exposures. Although toxicologists experimented with a variety of approaches, there emerged a scheme for assigning safe exposure levels that was based on the application of safety factors to experimental toxicity data, derived for the most part from studies in animals, but also from controlled studies involving humans(1). In general, toxicologists would divide experimentally-determined "no-observed effect levels" (NOELs) by such safety factors. The level of exposure arrived at by application of safety factors has never been claimed to be totally without risk, but it became widely accepted within the community of toxicologists that this type of scheme is appropriate for defining acceptable human exposure levels (except for carcinogens -- see below). Thus arose the

---

[1] Thus, one approach to deciding how much exposure to a carcinogen can be permitted is to set limits at whatever the detection capability of available analytical methods happens to be. The latter has, of course, no relationship to risk. This is not to say that analytical capabilities as well as a host of other factors should not play a role in decision-making. It is only to say that risk should not be ignored.

concepts of "acceptable daily intake" (ADI) for food and color additives and pesticides, and Permissible Exposure Limits (PELs) for exposures in the workplace(1,2).

The central concept underlying this approach is that for most forms of toxicity, the production of effects requires a certain minimum dose (a threshold dose), and that unless the minimum dose is exceeded, no effect will occur(2).

The experimental NOEL may approximate such a threshold dose in the small and relatively homogeneous group of test animals studied. However, there are plausible biological reasons as well as empirical evidence to show that the threshold dose is not fixed; that it varies, sometimes greatly, among individuals in a population; and that some members of the human population may be more susceptible than experimental animals to the toxic effects of chemicals. It thus became the practice to apply safety factors to NOELs in order to compensate for these possibilities, for the other scientific uncertainties described earlier, and for limitations in the quality of the experimental data(3).

This safety assessment scheme, which is still in wide use, has never been claimed to provide absolute safety (zero risk). There is, in fact, no scheme that could do so. But it does claim that any residual risk associated with exposures corresponding to an ADI is almost certainly very low(3). This is probably the case for most types of toxic agents, but we have no method to determine whether it is. But because the scheme claims to provide an estimate of low risk exposures, it is, at least implicitly, a risk assessment scheme that makes no attempt to characterize the risk that remains at exposures said to be "acceptable".

Limitations In The Safety Assessment Scheme

The safety assessment scheme described above appears to have provided adequate public health protection, and will no doubt continue in use for some time to come. There are, however, certain limitations in the scheme that should be acknowledged.

First, the use of ADIs (or their equivalent) tends to give the impression that exposures to chemicals are either "safe" (below the ADI) or "unsafe" (above the ADI). Those who work in the area know that this is a false interpretation, because risk to a population does not simply "disappear" at a given dose. In fact there may be for some agents a range of doses well above their ADIs that fall well within the low or even zero risk category. On the other hand, risk may sometimes rise rapidly through and above an ADI. The point is that there are no sharp divisions in the continuum of dose-risk relations, at least insofar as we are concerned with population, not individual, risks(3).

It should be recognized that, no matter what risk assessment scheme is used, there will finally emerge an exposure level which will be said to be acceptable. There will probably always be a tendency to view such "official levels" as the dividing lines between "safe" and "unsafe" exposures. We suggest, however, that the use of a scheme that provides explicit estimates of risk, and from which policy-makers decide on the risk that is tolerable in specific circumstances, is less likely to be misinterpreted as providing such sharp distinctions.

Procedures for estimating and using NOELs can be wasteful of data(3,4). The selection of the highest dose at which "no effect" is observed (the NOEL) ignores the possibility that the lack of observed effects could have been the result of chance variation about a true effect. If two experiments, identical except for sample size, yield identical NOELs, the larger experiment provides greater evidence that a true NOEL has been observed, and hence greater evidence of safety. The NOEL approach also does not fully utilize the experimental dose-response information. Dose-responses that decrease sharply with decreasing dose have different implications for risks at doses below the observed NOEL (i.e., the human dose) than do shallower dose responses. However, this difference may not be accounted for in the setting of ADIs.

Serious questions can also be raised about the use of specific "safety factors" to establish ADIs without scientific evidence to support the magnitude of such factors. In fact, there is nothing but custom to support the use of any specific safety factor(3,5). Because it can also be reasonably argued that the selection of specific safety factors is a matter of policy, not science, the safety assessment scheme can be seen as a blend of scientific and policy decisions that cannot be easily separated.

It appears, then, that some modification in the "NOEL-safety factor" approach is in order. There are difficulties that must be overcome before we can arrive at suitable alternative methods but it is time to begin to move away from the concept that toxicologists can decide what is "safe" by simply selecting arbitrary "safety factors". We need to find ways to use the dose-response information in establishing ADIs, and also to distinguish explicitly the scientific aspects of these types of analyses from the policy aspects.

Finally, the scheme has generally not been considered, even by its proponents, appropriate to apply to carcinogens. This view may stem from the legal stricture (which exists in the United States in the form of the Delaney clause of the Food, Drug and Cosmetic Act) that no ADI can be established for a carcinogenic additive, in which case no safety assessment scheme is needed. On the other hand, it may stem from a scientific view that the mode of action of carcinogens is such that exposure at a calculated ADI (experimental NOELs can be defined for many carcinogens) is almost assuredly going to pose a risk of cancer, regardless of the magnitude of the safety factor. Exposure to other types of toxic agents at a calculated ADI will, in many cases, also pose a finite risk. For both carcinogens and other types of toxicants, it is not possible to show rigorously that zero population risk is achieved at any finite dose. It is possible, however, to estimate low or even negligible risk doses for all forms of toxicants, including carcinogens, although we suggest that the traditional methods for establishing ADIs are probably not the best ways to accomplish these goals(4).

Newer Concepts Of Assessment

It is clear that not all chemicals that exhibit carcinogenic properties can simply be banished from our society. It has become necessary to establish a systematic means for deciding the extent to which human exposure to carcinogens should be limited. It was in this context that a distinctly different scheme was developed to establish acceptable exposures. In its idealized form, this scheme involves two major and distinct steps(6):

(1) Risk assessment is performed to determine the nature and magnitude of risk associated with various levels and conditions of human exposure to a carcinogen.

(2) Risk management analysis is performed to decide the magnitude of risk that is tolerable in specific circumstances (i.e., in the context of current statutes and various control options).

Under this scheme, a decision on acceptable exposures is made in the second step, and involves matters of policy quite distinct from those issues concerning the nature and magnitude of risk. Under this scheme, the role of the health scientist is far more restricted than it is in the traditional safety assessment described earlier. The health scientist is no longer responsible for assigning acceptable exposures. On the other hand, the scientist has a more demanding task than under the traditional scheme, because he or she is asked to make an explicit statement about risk.

This scheme appears to have a number of desirable features. Most of all, it requires recognition that science alone can not decide what is safe or acceptable(6). (It must be acknowledged that many scientists remain convinced that science can, indeed, make such decisions. We believe this is an incorrect view.) Further, it requires that health scientists focus more directly on the essential scientific problems of risk assessment and come to grips with all of those fundamental gaps in knowledge described in the opening section of this paper. Under this scheme, the role of scientists is thus to: 1) define the most rigorous and systematic approaches to assessing risk that can now be found and justified, taking care to describe all the uncertainties attendant upon this task, so that some statement can be made about risk; and 2) conduct the research necessary to reduce these uncertainties. In other words, the role of the health scientist is to measure risk and also to describe and improve methods to predict risks under conditions of exposure for which risk information can not be directly collected.

Risk Assessment

Under the definition of risk assessment we propose, it is a broad activity, by no means limited to the uncomfortable problem of high-to-low dose extrapolation, which many people take it to be(6). It includes, as its first step, the problem of hazard identification and evaluation. In brief, this problem involves review and evaluation of various types of experimental and epidemiological information for purposes of identifying the nature of the hazards associated with a substance or activity. It is designed to answer questions such as: Is (substance x) a carcinogen? What type of carcinogen is it? What is the likelihood that the experimentally observed carcinogenic response is somehow uniquely related to the conditions of experimental exposure? What is the nature and strength of the evidence supporting this evaluation? The successful execution of this step depends on a fundamental belief in the unity of biology, but is also dependent upon a realization that interspecies differences in response are always possible and need to be considered.

The second step, termed dose-response evaluation, involves identifying the observed quantitative relationship between exposure and risk, and extrapolating from the conditions of exposure for which data exist to other conditions of interest(6). This step almost always involves high-to-low dose extrapolation and frequently involves extrapolation from experimental animals to humans. This step requires the assumption that dose-response relations do not simply disappear at the detection limit of our experimental or epidemiologic systems. It also requires that a biologically plausible mathematical function be employed to characterize the low end of the dose-response curve. Finally, it frequently requires the imposition of assumptions regarding the quantitative relationship between test animal dose-response functions and those expected to apply to human populations.

The third step is identification of the conditions of exposure (broadly defined to include intensity, frequency and duration) of the human population group that might be at risk and for which protection is sought(6). The last step involves combining the information on dose-response with that on exposure to derive estimates of the probability that the hazards associated with a substance or activity will be realized under the conditions of exposure experienced by the population group of interest. Risk assessment involves integration of the information and analysis associated with these four steps to provide a complete characterization of the nature and magnitude of risk and the degree of confidence associated with this characterization. A critical component of the assessment is a full elucidation of the uncertainties associated with each of the major steps(6).

Under this broad concept of risk assessment are encompassed all of the essential problems of toxicology that traditional safety assessment schemes have dealt with, but they have been recast to provide a means for answering a different question -- that is, the question of risk. There are other important differences as well. Risk assessment does not rely on the biologically and statistically dubious concept of a NOEL, but takes into account all of the available dose-response data. It treats uncertainty not by the application of arbitrary safety factors, but by stating them in qualitatively and quantitatively explicit terms, so that they are not hidden from decision-makers. Risk assessment defined in this broad way forces an assessor to confront all the scientific uncertainties and to set forth in explicit terms the means used in specific cases to deal with these uncertainties. And, of course, risk assessment does not include those decision-making processes necessary to establish acceptable exposure conditions.

Risk Management

Completion of a risk assessment yields no view of whether the projected risks are important and require the imposition of controls. We here enter the realm of risk management, which is far less well-developed than even the fragile domain of risk assessment(3,6). Some contend that risk management decisions are strictly matters of policy. We do not argue this point, but add that this does not mean they should be devoid of objective, analytic support. The problem seems to have two primary components. The first involves a decision on whether or not the assessed risk is important. This decision, we suggest, should not be based solely on the magnitude of the projected risk, but also on the degree of confidence that can be placed in both the data underlying the assessment and the methods and assumptions used. The degree of confidence is a function of several aspects of the assessment, including the strength of the evidence supporting the conclusion that a substance is indeed hazardous (e.g., that a chemical is a human carcinogen), the extent to which supporting data are biologically and statistically concordant, and the extent of variability in the risk when it is predicted under different assumptions and models. Some means is needed to permit systematic consideration of all of these types of information in the decision-making process, but little analytic work has yet been done in this area.

Some agencies have defined negligible or *de minimis* risk for some carcinogens strictly in quantitative terms(7). This approach may be a reasonable place to start analysis, but it fails to recognize that the data bases for different carcinogens vary widely in quality and content, and that several other non-quantifiable factors (that we include as part of the assessment of "degree of confidence") influence the risk. In other terms, two substances apparently posing the same quantitative risk may, in fact, produce quite different risks. We suggest that the other non-quantitative information available in the risk assessment can serve as a guide to determining the likelihood of such differences.

If it is decided that a risk is worth worrying about, additional analysis is needed to decide how and to what extent control is necessary. This area involves questions of cost, technical feasibility, and law, all of which we leave to others.

Conclusion

The safety assessment scheme now applied to toxic agents other than carcinogens could be modified so that better advantage is taken of dose-response information and so that scientific aspects of the scheme can be distinguished from the policy aspects. Decisions on appropriate safety factors, if needed, would be associated with the domain of policy-making, their magnitude depending on scientific judgments regarding uncertainties in the data and dose-response relations.

As currently practiced risk assessment is conceptually sound, but the uncertainties are great because of gaps in fundamental knowledge. Research into underlying mechanisms of toxicity, as they bear on knowledge of dose-response relations at low dose, is critical to further advances in this field. Clearly the highly insensitive research tools we now have cannot be relied upon indefinitely as the basis for these important public health decisions.

Literature Cited

1. Food Safety Council. *Food Cosmet. Toxicol.* 1980. 18, 711-734.

2. Oser, B.L. *Arch. Environ. Health.* 1971. 22, 696-698.

3. National Academy of Sciences. *Drinking Water and Health.* Washington, D.C. 1977. pp. 22-55.

4. Schneiderman, M.A. J. Wash. Acad. Sci. 1974. 64, 68-78.

5. Weil, C.S. Toxicol. Appld. Pharmacol. 1972. 21, 454-463.

6. National Research Council. Risk Assessment in the Federal Government: Managing the Process. National Academy Press. Washington, D.C. 1983.

7. Food and Drug Administration. Fed. Reg. 1979. 44, 17070-17114

RECEIVED November 4, 1983

2

Use of Toxicity Test Data in the Estimation of Risks to Human Health

NORTON NELSON

Institute of Environmental Medicine, New York University Medical Center, New York, NY 10016

Historically there has been an enormous elaboration of techniques for evaluation of the toxicity of chemicals in the last thirty years. At that time chronic lifetime tests in rodents were just coming into application and tests on human subjects, prisoners and "volunteers" were not infrequent. On the other hand, there have been perhaps some retrograde changes, namely in the less frequent use of some of the larger species, such as cats, rabbits, dogs and primates. It is perhaps also true that there is now greater routinization than in earlier decades with somewhat less attention to fitting the toxicity test to the chemical and to the circumstances.
The basic problems remain: biological transfer from one species to another and the need for better quantitation, greater sensitivity, and higher efficiency in cost and time. Larger test groups have brought some improvement in quantitation and sensitivity. The use of human subjects has virtually and properly disappeared with growing concern for the ethical issues involved. A heavy preoccupation with cancer as the endpoint has in some degree lessened interest in other sometimes more important endpoints. Hopefully this trend will be reversed under the new National Toxicology Program which will attempt to broaden the range of information secured.
We have had many attempts to develop short term tests aimed at securing the needed information in a shorter time and less expensively. The bacterial revertant test is clearly outstanding in this regard. This still has defects which may be amenable to correction. Improvement in fields other than mutagenesis (and cancer) has been extremely uneven, and there is no counterpart "success story." An objective of the

0097-6156/84/0239-0013$06.00/0

future is to expand the range of short term tests and reduce the need for whole animal studies. Means for using all dose points for estimating a "pseudo" no observed effect level (NOEL) is suggested.

It may well be, however, that major improvement is not to be sought in finding more rapid models for existing toxicity tests but to develop a synthesis of independent information acquired by ancillary routes. Thus, the Ames system is limited in the sense that it fails to deal, for example, with mammalian repair mechanisms. Such information might be specifically sought in separate tests. Similarly, the pharmacokinetic aspects of movement from point of entry into the body to the target tissue and target biochemical unit (DNA) could expand the utility of simple tests.

In more general terms and with endpoints other than cancer, one can visualize the synthetic assembly of information from a variety of studies which could inform as to some of the biological factors that we know are involved and which cannot be derived from a single test; the tissues obtained through surgical operations and autopsies could supply the needed human tissue. It seems possible that such an approach applied to a variety of endpoints could strengthen very substantially both the quantitative and qualitative aspects of toxicological assessments and could, therefore, make quantitative risk assessment more meaningful.

These and other opportunities to improve and make more efficient toxicological appraisals for risk assessment will be discussed.

The last 30 years have seen major changes in the practice of toxicology, both qualitatively and quantitatively. Quantitatively the conduct of toxicological pretesting has expanded very substantially; there are now many contract laboratories available for the conduct of such work. Qualitatively many changes have occurred over that period of time. Chronic lifetime testing, especially with the cancer endpoint in mind, was already established but was relatively new as a regular part of toxicological pretesting. Indeed one of the great triumphs of toxicological testing was the identification of the carcinogenicity of AAF in 1941 by Wilson, et al. (1). This compound, which was originally proposed as a pesticide, was found to be carcinogenic in those FDA tests, thus aborting its use as a pesticide but, at the same time, providing the experimental cancer community with one of the most widely used research carcinogens.

At that time, test groups were generally much smaller than they are now, and a fuller awareness of the importance of group size and the standardization of test procedures has developed. It is also perhaps true that there has been a major trend towards routinization in toxicological tests; in some cases this is a step in the wrong direction, since routinization brings with it two dangers: one that tests irrelevant to the chemical or to the expected use may be undertaken, and the other is that a thoughtful specific adaptation of the test procedures to the particulars of the circumstances may be omitted. Thus, unneeded things may be done and needed things may not be done. There is another danger of overstandardization in the sense that when a producer of a chemical is given precise instructions as to what tests are to be conducted, he is to some degree relieved of the intellectual and ethical responsibility for using the best available science and art to establish the safety of the compound for the proposed uses. In this sense, the one who conducts the test may say he has supplied the information requested and so has fulfilled the legal commitment. The petitioner is thus freed of any implication of responsibility for exercising his own ingenuity and scientific acumen in using the best of the available science to establish the safety of the agent in question.

There are, of course, other reasons for standardization, especially in regard to cross-comparison and historical comparability of data, but the dangers in over routinization are ever present, should be recognized, and constantly questioned by the responsible toxicologist.

There have been perhaps some other retrograde movements, such as a lesser tendency at the present time to use larger species, such as rabbits, cats, dogs, monkeys, and to depend almost exclusively on rodents. Perhaps to some degree these are inevitable prices to pay for the more widespread use of toxicity prescreening tests which in itself is of course a most salutary trend. The basic problems in the field remain, that is, the uncertainties in transferring data from the test species to man, the need for better quantification, greater sensitivity, and higher efficiency in cost and time. Although the sensitivity of toxicological tests has improved somewhat with the trend toward larger group size, the sensitivity thus achieved is in many cases far short of that relevant for direct transfer of the findings to man. In the case of cancer, for example, incidence rates as a minimum applicable to man of the order of 10^{-4}, 10^{-5}, or 10^{-6} are imperative; this is, of course, not even remotely achievable in practical laboratory experiments. In some degree transferability of data may have been impaired through the reduction in the number of species generally used; in addition, it must be kept in mind that the trend toward use of highly inbred strains (although desirable from the point of view of uniformity of response) nevertheless leads towards the use of test animals with highly specific susceptibilities which may lead to missing other endpoints were an outbred strain with more genetic diversity used.

The use of so-called "no-effect" levels in estimating "safe" levels for man from laboratory studies has a long tradition. In 1975 the author called attention to the limited value and statistical meaninglessness of this term, especially when the group size is not specified; at that time he suggested that the term should at least include the qualifier "observed," that is, the no-observed effect level (NOEL)(2). This term is at least more accurate; however, it still does not normally make full use of dose response data. As presently used, the NOEL is defined as a point between two sequential data levels, one with an observed effect and one with none. Thus, it essentially represents the use of a single point in the positive dose response data. This sometimes involves discarding significant additional data. I would propose that an alternate technique be used to develop what I would tentatively call a "pseudo" NOEL. This would involve fitting a curve to the observed dose response points. Any one of a number of procedures could be used here (e.g., the probit or logit curve). One would then find the dose level corresponding to an arbitrarily selected low incidence point, e.g., 1%. One percent is an incidence level which could be easily overlooked in most laboratory studies in a single experiment using 50 animals or even in several experiments. This 1% limit would be regarded as a "pseudo" NOEL; it would perhaps sometimes correspond to an actual NOEL. This technique would permit use of all data in the selection of this starting point for whatever subsequent data treatment is desired. It would bring with it such statistical parameters as confidence levels. Thus, one would replace the present NOEL with an artificial one based on an arbitrary incidence level that corresponds to that incidence level which may or may not be detectable in normal experiments with groups of 50 animals.

What one then does with a NOEL would require further consideration. One could use this "pseudo" NOEL with a safety factor or one could use it as a point for a linear extrapolation to "0" (or the background level) for example.

A somewhat similar approach for a different purpose has recently been proposed (3).

It may well be that a more competent statistican than the author will choose other intercepts or other techniques. However, the basic objective is to use all positive data and in a manner which will permit the development of confidence levels.

The use of human subjects for test purposes, including studies on prisoners once widely used, has happily essentially disappeared. It has not, however, been adequately replaced by careful study of individuals who have already been exposed to toxic agents; thus the wider use of clinical follow-up and biological monitoring is an urgent need in this issue of transfer of information from the test species to man.

A growing emphasis on cancer as the endpoint has in some

degree preempted interest from other equally important endpoints. It now appears that this trend may be stemmed and perhaps reversed with the development of the new National Toxicology Program which will systematically work towards the development of tests aimed at revealing effects other than cancer, such as on the various organ systems and behavioral responses.

The drive towards securing information less expensively and in a shorter time has met with outstanding success in cancer testing where the bacterial revertant tests, such as the Ames Test, have proven to be a very useful screen for mutagenic agents and, thus, for certain kinds of chemical carcinogens. Similarly, cell transformation studies and tests for DNA damage have been developed which can strengthen the relevance of tests of this sort for potential carcinogenicity and mutagenicity. In other fields of toxicology, there has been substantially less success. A number of attempts have been made in the field of teratology, and these have some utility; but by and large attempts to develop short term tests using isolated cell or enzyme systems have not been highly fruitful. Again, although some usage of organ function tests is underway (particularly lung, liver and kidney), these have not been systematically explored with the view toward adapting them efficiently and meaningfully to laboratory animal studies. What is required here is a systematic attempt to streamline such tests and to improve them in respect to sensitivity, repeatability and informativeness. This is an area in which some degree of standardization would be highly useful and represents a field for systematic study. Of course, the need is not merely to shorten the time of testing and save money, but to improve their utility.

An interesting formal treatment of risk estimation has recently been put forward by Nordberg and Strangert (4). Conceptually it deals with compartmental movement, metabolism and the definition of the critical organ, critical effects and critical concentration. It also defines a new concept, the "damage" function," relating to the critical injury.

There is a strategy which has been used only to a limited extent which merits full scale exploration as a route to better information, better quantitation and greater relevance to humans. I refer to the orderly assembly of information from different tests into a coherent approach to an attempt to reliably relate laboratory data qualitatively and quantitatively to human health effects. Figure 1 illustrates some of the routes and mechanisms which determine the end effect of a toxic chemical on a mammalian organism, be it man or a rodent. This is intended to illustrate those steps which each of us, of course, are very familiar with, namely entry into the body via inhalation, via skin penetration, via oral ingestion, the extent of absorption, alteration during or after absorption, through enzymatic or chemical processes (toxication-activation, detoxication-inactivation), the transport through the organism (rates depending upon compartment interfaces, whether the process is active or passive), the attack on the end

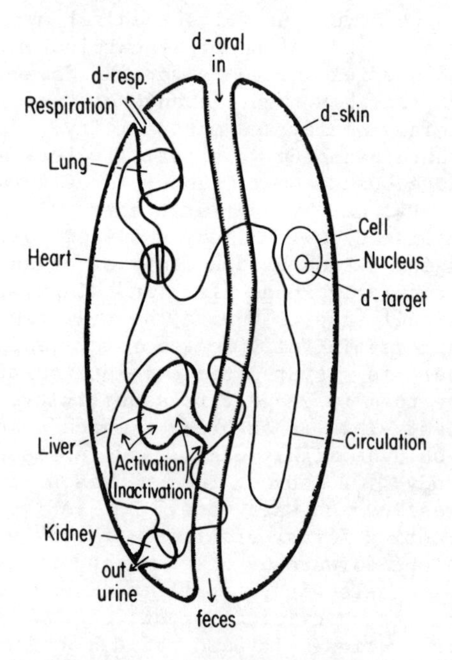

Figure 1. A Mammal (slightly simplified).

critical cell or biochemical unit, and finally the repair mechanisms (or, if function is altered, the restorative functional capacity). Even this complicated sequence of issues is, of course, a simplification; nevertheless, it does represent a pattern which in varying degrees determines the outcome. The strategy suggested here is to first define the critical processes or critical organ. This could be done where possible on structural grounds, by analogy with other chemicals, or best from laboratory assays of appropriate length and complexity. The next step would be to undertake parallel laboratory studies of animal and human (surgical, autopsy) tissues to establish the qualitative and quantitative relationship between the test species(s) and humans. Having defined the critical organ, cell or biochemical unit, the objective would then be to define the relationship between the entry dose (e.g., inhaled, ingested, etc.) and the target receptor dose through these ancillary studies. This relationship obviously involves the pharmacokinetic and metabolic patterns to which the chemical is subject. Next to be taken into account are the nature of injury to the target system, the repair (or functional adaptive response) and the reversibility of the effects. Such an approach would involve, according to need, study of isolated systems (human tissues as well as animal tissues), pharmacokinetic studies (on laboratory animals), and the examination of repair mechanisms. Examination of metabolic activation or inactivation will involve organ systems, isolated enzyme or cell systems including cell cultures, as required.

The strategy is then to attempt to identify the particularly critical stages between exposure and effect and to focus study on these in a comparative manner. The attempt would then be to synthesize or assemble these components into a quantitative and qualitative chain linking the laboratory studies to man.

Such an approach is outlined in skeleton manner in Figure 2, which in a very much simplified manner suggests an organizing scheme for animal to man extrapolation of chemical carcinogens. I wish to acknowledge my indebtedness to my colleague, Professor Bernard Altshuler, for this schema. As you will note, it briefly outlines the several stages of entry, activation, inactivation, movement to the target biochemical unit (DNA), on to the several repair mechanisms, initiation, early cell transformation, cell progression and growth (frequently through a benign stage), finally to uncontrolled growth and a malignant tumor.

The strategy proposed here would very much depend on the use of human tissues from accident cases, from surgical operations, and such sources; the objective is a qualitative and quantitative comparison of human tissue with the tissues of the species(s) studied in the laboratory.

At this time our biological knowledge of the action of chemical carcinogens makes the application of this strategy to cancer particularly appealing. Even so, it has not yet been applied in a systematic manner. There have been a series of

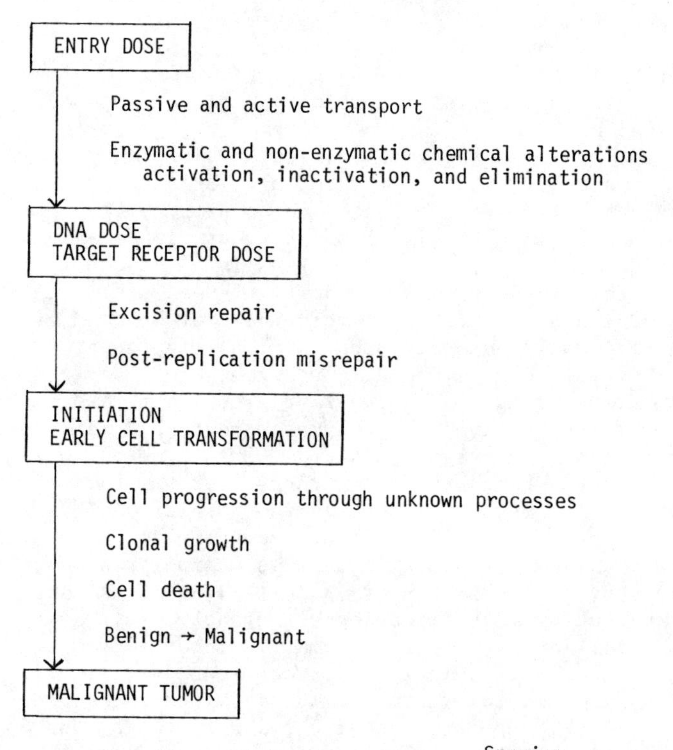

Experimental	Species (Strain, Sex)	Organ
Animal Experiments	Man	Skin
Cell Culture: Mutagens, transformation	Mouse	Liver
	Rat	Bladder
Organ Culture: Inflammation, other lesions	Hamster	Lung
	Possibly others	

CARCINOGENS

Direct Acting: BCME, BPL, DMCC, EPI
Indirect Acting: FANF, Nitrosomines, AAF, Aromatic Hydrocarbons
Promoters
Cocarcinogens
Methotrexate

Figure 2. Animal-to-man extrapolation: organizing scheme.

isolated studies which use components for this purpose, such as the work of Gehring and colleagues (5,6) on the examination of the pharmacokinetics of vinyl chloride, their importance for dose-response relationships of vinyl chloride, and their implications for man; these were also explored by Anderson, Hoel and Kaplan (7). Studies by Autrup, et al. (8), examine comparative patterns of tissue metabolism of polynuclear aromatic compounds. Direct parallel cross-species studies of repair mechanisms of damaged DNA relevant to this strategy will, of course, also be needed. This is obviously a critical issue in moving from such simplified systems as bacterial revertant tests to mammalian systems where repair mechanisms are of vital importance and are very different in bacterial than in mammalian systems.

The relatively orderly issue of extrapolation of cancer from laboratory to man, although very complex, is nevertheless probably closer than other non-cancer endpoints to providing underlying concepts upon which to develop this "synthetic" risk assessment; the full development of this strategy to other endpoints may be well in the future.

Nevertheless, I believe that it is only through such attempts, carefully and selectively applied, that we will move beyond the present long term elaborate, expensive and poorly informative toxicological studies toward an approach that may be more reliable, more quantitative and more relevant to man, perhaps in some cases shorter in time and even perhaps less expensive. Neither of the last two objectives should, however, be of overriding consequence.

Quantitative risk assessment depends on data that are reliable, sensitive and quantitative. It may well be that the numerical extrapolation from the current small scale (but manageable) laboratory tests can be substantially improved and moved downward to the effects of lower dose levels through the shrewd use of these isolated cell and biochemical test systems where the interplay of inactivation, activation and target molecule injury can be studied at concentrations well below those possible where one is looking at endpoints in relatively small groups of whole animals.

Although I have dealt in broad generalities and no doubt have simplified many issues and underestimated the scientific difficulties, nevertheless the promise of such strategies is so great that substantial endeavors in selected areas should be undertaken now without further delay. Unquestionably errors will be made and false starts will ensue, but this is inevitable in dealing with a field of this degree of complexity.

Literature Cited

1. Wilson, R.H.; De Eds, F.; Cox, A.J. Jr. Cancer Res. 1941, 1, 595-608.
2. Nelson, N., Chairman, Committee for the Working Conference on Principles of Protocols for Evaluating Chemicals in the Environment; National Academy of Sciences: Washington, D.C., 1975, 454 pp.
3. Albert, R.E., personal communication.
4. Nordberg, G.F.; Strangert, P. "Risk Estimation Models Derived from Metabolic and Damage Parameter Variation in a Population," 1982, to be published.
5. Gehring, P.J.; Watanabe, P.G.; Blau, G.E. Ann. N.Y. Acad. Sci. 1979, 329, 137-52.
6. Ramsey, J.C.; Gehring, P.J. in Health Risk Analysis, Proceedings of the Third Life Sciences Symposium, Gatlinburg, TN, 27-30 October 1980; Richmond, C.R.; Walsh, P.J.; Copenhaver, E.D., Eds.; Chapter 17.
7. Anderson, M.W.; Hoel, D.G.; Kaplan, N.L. Toxicol. Appl. Pharmacol. 1980, 55, 154-61.
8. Autrup, H.; Wefald, F.C.; Jeffrey, A.M.; Tate, H.; Schwartz, R.D.; Trump, B.F.; Harris, C.C. Int. J. Cancer 1980, 25, 293-300.

RECEIVED July 5, 1983

3

Interspecies Extrapolation

DANIEL B. MENZEL and ELAINE D. SMOLKO

Departments of Pharmacology and Medicine and Comprehensive Cancer Center, Duke University Medical Center, Durham, NC 27710

Animal experimentation produces most available data for chemical toxicity. Methods for using this data in assessing human risk are presented, with emphasis on mathematical modeling. Any interspecies extrapolation effort must account for variations in morphology and metabolism. Provided a general similarity exists, the specific differences do not preclude analysis. Application of a mathematical model using anatomical, rather than pharmacokinetic, compartments for determination of toxicity of chemicals is discussed. The Miller Model is presented as a method for quantitative assessment of tissue dose of toxicant following inhalation. Metabolism is discussed in terms of reactive intermediates and of species and strain variations. These approaches indicate progress in the use of animal toxicology data for predicting human risk.

Chemical threats to human health dictate a careful appraisal of new chemicals. A continued reappraisal of known toxicants is also needed to ensure that the human health risks are balanced by benefits from the use of these compounds. The toxicity of chemicals is largely determined by animal experimentation. The risk to man is estimated by interspecies extrapolation from animals to man.

The basis for animal experimentation is the presumed similarity between animals and man. This assumption is so commonplace that it has become a truism. Yet, the specific differences between man and animals become more apparent as quantitative and precise measurements of toxicity become increasingly available. Are animals good surrogates for humans? Do animal experiments present an accurate picture of the hazards to man of chemical exposures? Can animal experiments be

0097-6156/84/0239-0023$06.00/0

used to predict quantitatively the outcome in man? Do lifetime exposures of animals present an analogy with human lifetime exposures? These are but a few of the questions raised daily in the conceptualization of animal experiments and use of resultant data in societal decisions. In a certain sense, these are philosophical questions; but in another sense they are highly practical, and solutions are urgently needed.

We will discuss some recent approaches to these questions. Our remarks will be restricted to chemicals and to interspecies extrapolation. The aim of this discussion is to provide a framework for increasing the precision of experiments using animals as surrogates for man.

Interspecies Differences in Morphology

The morphology of animals is so apparently different from that of man that it is often overlooked in the interpretation of test results or in the selection of appropriate species for testing. Comparative anatomical studies have revealed important similarities as well as dissimilarities. Inhalation toxicology experiments, for example, are particularly sensitive to anatomical differences. Quantitative morphometric studies of the human and animal lung were begun by Weibel (1) , who used a specialized statistical method to sample the highly heterogeneous structure of the normal lung. These studies and those of Kliment (2) led to an anatomical model which describes the equally bifurcating nature of the human lung. Figure 1 is a schematic representation of these relationships between the tube diameter and length, and the number of bifurcations. Each bifurcation is referred to as a generation. The number of generations in animal lungs differs from that in human lungs, mainly because of the smaller size of animal lungs compared to those of adult humans. Also, rodent lungs differ in the generation at which alveoli begin to appear branching off from the main bronchi or breathing tubes. The alveoli represent the gas-exchange regions of the lung and are important sites of uptake of inhaled toxicants. Detailed morphometric analyses of rat, guinea pig, and rabbit lungs have been reported. Studies of mouse lungs are now in progress. These data, combined with continuing studies of the human lung, will provide a "map" of the lung showing its dimensions with relation to the number of generations. As discussed below, such a map can be described mathematically and used in a model of the regional deposition of gases and particles in the lung.

While human and animal lungs are dissimilar in size and number of generations, they are strikingly similar in their manner of organization. Variations in details have been noted and measured, including such features as angles between bifurcations, size, and thickness of tube and alveolar wall. These distinctions are, however, amenable to analysis and

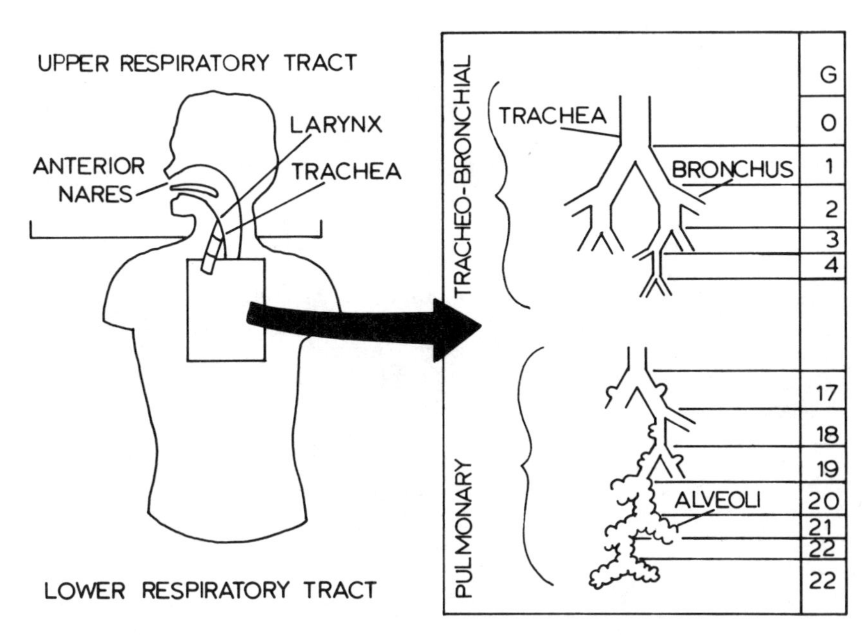

Figure 1. Schematic representation of relationships between tube diameter and length and number of bifurcations in the human lung.

extrapolation through physical principles of gas flow and aerodynamics of particles in gases. Differences between human and animal lungs can be turned to advantage once quantitated, provided a general similarity exists.

The deposition of gases and particles in the nasopharyngeal region of the respiratory tract is likely to be of industrial importance, since the work place is often contaminated with relatively large particles likely to be deposited in the nasopharynx and not in the lung. Recently, inhalation studies of formaldehyde spurred comparative studies of the nasopharyngeal region of the respiratory tract. Mice and rats developed nasal tumors when exposed to levels of formaldehyde near those occurring in the work place. Similar tumors have been reported in workers exposed to formaldehyde vapors. Workers in nickel refineries have an increased incidence of nasal tumors, presumably because of the deposition of nickel aerosols in the nasopharynx. Schreider and Raabe (3) examined three species of animals by producing silicon rubber casts of the nasopharynx. These casts of dogs, rabbits, and monkeys revealed a highly complex, convoluted pathway leading to the lungs. Sections through these casts were made, and the area as a function of the distance from the exterior to the interior was compiled. By combining the measured areas with the air flow through the nose, the Reynold's number can be computed to indicate the turbulence of the air flowing through the nasopharynx during breathing. Such calculations lend themselves to predictions of the deposition of aerosols within given regions of the nose. The naospharyngeal removal of gases can be measured directly (4), but these measurements are difficult to make and are necessarily restricted to a few values of flow. An anatomical description in mathematical terms, on the other hand, allows a more general approach. Gas uptake can be modeled in terms of the physical properties of the gas and the gas uptake in physiological fluids, as described below for the lung.

The diversity in the nasopharynx of rodents and man makes rodents less useful for studies of toxicity of large particles or toxicants readily removed by solution. Rodents are required to breathe through their noses. Major differences in dose and dose-rate are likely, then, between man and rodents for compounds deposited predominantly in the nasopharynx.

The rat, but not the hamster, mouse, rabbit, and guinea pig, has mucous glands as does man. Lamb and Reid used the rat to produce experimental bronchitis from inhalation of sulfur dioxide and cigarette smoke (5-7). It is questionable if other animal species would have responded similarly, because of the anatomical differences.

General Mathematical Models in Toxicology

Considerable progress has been made in applying pharmacokinetic modeling to animal data and extrapolation to man. These models seize upon the similarities and dissimilarities between species. Himmelstein and Lutz (8) suggest that models built on "physiological pharmacokinetic" principles can confidently predict effects in man. These models use basic physiological and biochemical information to develop differential equations describing drug or toxicant distribution and deposition. These models are characterized by anatomical (organ volumes and tissue sizes), physiological (blood flow rate and enzymic reaction rates), thermodynamic (binding isotherms), and transport (membrane permeability) considerations. A rational mathematical model also aids in the direction of research and testing of hypotheses which are sometimes difficult or impossible to test directly.

As an example of the application of this methodology. Dedrick and his associates examined the pharmacokinetics of the cancer chemotherapeutic drug, methotrexate (8-12). This physiological "scale-up" pharmacokinetics focuses on interspecies differences in size and perfusion characteristics of anatomical compartments rather than pharmacokinetic compartments. The physiological parameters and the set of differential equations that allow such prediction of plasma and tissue concentrations in man based on the data obtained in animals, at a given level and frequency of exposure, have been reported. This approach has been used successfully to adjust the dose of methotrexate used clinically to avoid undesired toxic side effects from the drug.

Application of Mathematical Models to Inhalation Toxicology

Because the lung is composed of over 40 different cell types which are regionally concentrated, knowledge of the regional dose of a toxicant to the lung is very important. Inhaled gases may affect only the upper, middle, or lower respiratory tract. The symptoms resulting from such regional distribution are quite distinct. For example, sulfur dioxide exposure results predominantly in chronic bronchitis in rats (5), while chronic exposure to ozone or nitrogen dioxide leads predominantly to emphysema (13). Chronic bronchitis is restricted to the upper airways, stimulating the production of mucus and obstruction of the major airways; emphysema is restricted to the respiratory region of the lung and decreases gas exchange by decompartmentalization of the alveolar region of the lower respiratory tract. At present, direct measurement of the regional dose of an inhaled toxicant is difficult, if not impossible. An alternate approach is to combine the anatomical models of the lung with the physical properties of the inhaled gas and its chemical reactivity with cellular constituents and products to predict which regions of the lung are most likely to receive the greatest dose; that is, to provide a specialized model based mostly on anatomical features of the lung relevant to regional uptake of toxicant.

Using the bifurcating model of the human lung and morphometric data on guinea pig and rabbit lungs, Miller, et al. (4) demonstrated the similarity between animals and man in regional pulmonary deposition of ozone (O_3). The transport and removal of O_3 in the lung was simulated by using a binary convective-diffusion equation:

$$\frac{\partial \bar{C}}{\partial t} + \bar{U}_x \frac{\partial \bar{C}}{\partial x} = (D_{mol} + D_{ed}) \left(\frac{\partial^2 \bar{C}}{\partial r^2} + \frac{1}{r}\frac{\partial \bar{C}}{\partial r} + \frac{\partial^2 \bar{C}}{\partial x^2}\right) + \bar{S}$$

where C, U_x and S^* represent species-averaged population concentrations, velocity, and source terms, respectively, in a given airway at a specified location and time. The axial and radial directions are x and r; t equals time; D_{mol} is the molecular diffusion coefficient of O_3; and D_{ed} represents the diffusion coefficient due to eddy dispersion. This equation represents a statement that the removal of O_3 by the lung is a function of convection, axial and radial diffusion, and chemical reactions.

Chemical reactions are assumed to occur instantaneously. Compared to the mechanics of breathing, the chemical rates of reaction of O_3 with cellular constituents and exudates are so fast as to be instantaneous. Thus, O_3 and the cellular constituents or exudates (mucus, in most cases) can not coexist in the same solution. The liquid phase can be thought of as consisting of two layers (14). The tissue dose, then, can be calculated from the case where the O_3 concentration in the overlying layer exceeds the concentration of the reactants secreted by the cell. In most parts of the lung, cells are covered with a mucus layer; from the chemical composition of the mucus and the stoichiometry of reaction of O_3 with these constituents, the dose of O_3 reaching the underlying cells can be calculated knowing the inhaled O_3 concentration. In Fig. 2, taken from Miller, et al. (4), the tissue dose of O_3 is plotted against the region of the lung for several inhaled O_3 concentrations. Remarkably similar plots were obtained for rabbit and guinea pig lungs. Even more important, the region of the lung receiving the largest predicted dose of O_3 is that which shows the greatest anatomical damage in actual exposures of animals (15,16). This region of the respiratory bronchiole and the alveolus was thought to be extraordinarily sensitive to

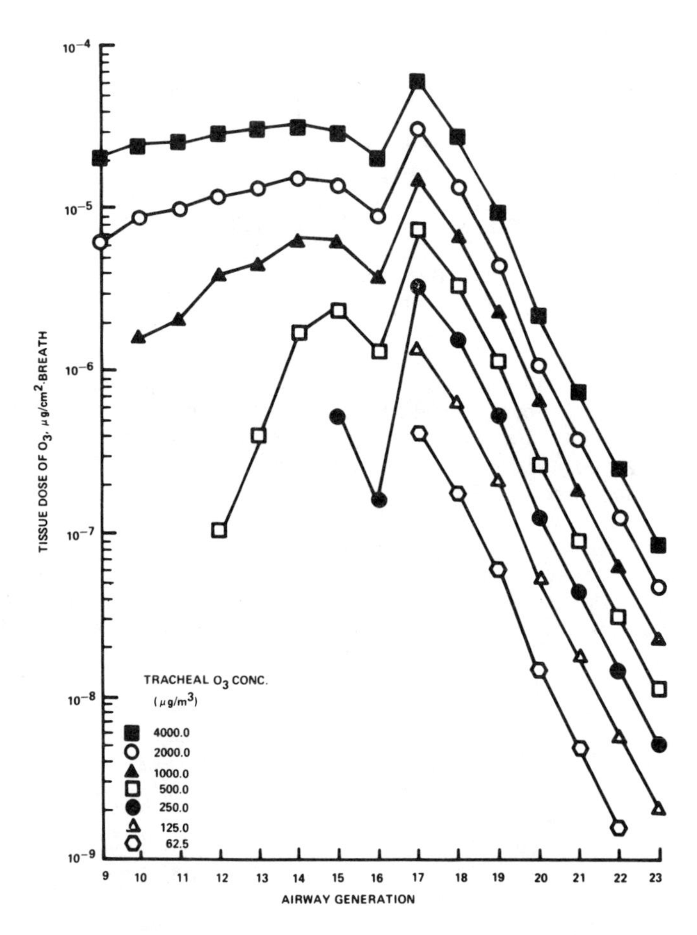

Figure 2. Tissue dose of O_3 plotted against the region of the human lung for several inhaled O_3 concentrations.

O_3, but these data suggest that the apparent anomalous sensitivity is really due to a difference in dose-rate.

When combined with measurements of the removal of O_3 from the nasopharyngeal cavity, quantitative estimates of the integrated tissue dose can be made. These estimates suggest that rabbits receive about twice the tissue dose of man for the same inhaled O_3 concentration. While regional similarities exist for man and these two animal species, quantitative dissimilarities are present. A study now in progress extends this approach to rats and mice, which have available a much larger compilation on the health effects of O_3. The scaling of these health effects to man at ambient concentrations of O_3 is also underway.

Polymorphic Xenobiotic Compound Metabolism In Animals and Man

Current thought holds that most toxic organic compounds and carcinogens are non-toxic or non-carcinogenic in their original form and must be metabolized to a more reactive metabolite or ultimate toxicant (17). This idea of "reactive intermediates" has been one of the most useful concepts in explaining toxicity of a number of compounds and has advanced considerably our understanding of the chemistry of toxicity and carcinogenicity. Most compounds which are converted to more toxic reactive intermediates are substrates for the mixed function oxidases (MFO), which are dependent on cytochrome P-450 (P-450) isoenzymes for activity. Depending upon the species and organ, as many as seven P-450 isoenzymes have been reported. P-450 isoenzymes are under genetic control in both man and animals. Using the antihypertensive drug debrisoquine, Smith and his colleagues have studied the genetic variations of several human populations and several species of rodents and strains of rats.

Debrisoquine is metabolized almost exclusively to 4-hydroxy debrisoquine (18). 4-Hydroxy debrisoquine and its parent compound are easily detected in the urine by gas chromatography. Urine is collected for 8 hrs following the oral administration of a single 10 mg dose of the drug. The ratio between drug and metabolite excreted in the urine ranges from 0.01 to 200. In man, the phenotype corresponding to extensive metabolizers (EM) ranged from 0.01 to 9, while poor metabolizers (PM) ranged from 20 to 200 (18). In a survey of 258 unrelated white British subjects, 8.9% were found to be the PM phenotype. The EM phenotype was dominant, and the degree of dominance was estimated at 30%. From studies of nine pedigrees, the PM phenotype was found to be an autosomal Mendelian recessive characteristic. These studies confirm and extend the previous estimates of PM occurrence of 6% in whites (19), 7% in blacks, and 1% in Egyptians (20). PM excrete only 1-3% of the drug and attain much higher blood levels than EM.

Other drugs whose metabolism by man is under the same genetic control as debrisoquine are guanoxon and phenacetin (21), phenytoin (22), metiamide (23) and 4-methoxyamphetamine (24). Antipyrine metabolism is, however, not under the same control as debrisoquine metabolism, despite the similarities of metabolism of these two drugs by the MFO system (25).

Diversity in oxidative drug metabolism has been demonstrated for 4-hydroxy amphetamine for the guinea pig and rat (24). The human EM phenotype excretes 4-hydroxyamphetamine primarily as the O-demethylated product, with minor amounts of parent drug, N-oxidation or b-oxidation products. The human PM phenotype excretes less overall drug; a large fraction is unchanged drug and N-oxidation product, with only small amounts of O-demethylated drug. Guinea pigs excrete the O-demethylated product exclusively and in large amounts. Rats excrete primarily the O-demethylated product, with some parent drug and N-oxidation product. Thus, the rat and guinea pig represent the human EM, but not the PM, phenotype.

Polymorphism in debrisoquine metabolism was demonstrated for the rat (26). Seven strains of rats were examined for their ability to metabolize debrisoquine. The Lewis strain was an EM, while the DA strain was a PM. Aside from the 4-hydroxy metabolite, rats also excreted 6-hydroxy debrisoquine. The DA strain excreted less of both metabolites. The Lewis and DA strains showed good recovery of the drug in 24 hr urines with 74.6 and 56% of the dose excreted, respectively. Phenacetin was used to test further the polymorphic nature of drug metabolism in these two strains, since the O-demethylation of phenacetin is under the control of the same gene locus as debrisoquine in man (21). Considerably less paracetamol was excreted by DA rats (38%) than by Lewis rats (54%). DA rats also had elevated levels of 2-hydroxy drug, a pathway associated with hemotoxicity in man (27).

Speilberg (28) recently reviewed the importance of genetic control of drug metabolism in chemical teratogenesis. Phelan, et al. (29) reported discordant expression of fetal hydantoin syndrome in heteropaternal dizygotic human twins. They suggest that the difference in response to hydantoin teratogenesis in man is due to differences in inherited ability to metabolize drugs. Speilberg cites experimental evidence in mice in support of this hypothesis. The Ah locus in mice, which enables induction of arylhydrocarbon hydroxylase, was manipulated by Shum, et al. (30) to demonstrate greater teratogenic risk in those fetuses possessing the Ah+ phenotype. Speilberg also points out the importance of the mother's phenotype in determining the blood concentration of the teratogen and, therefore, the transport of the chemical across the placenta to the fetus. In Speilberg's opinion, the uncertainty in current tests is too great to be of much help in patient counseling after drug or toxicant exposure. The alternative is a drug nihilism, as the result of physician uncertainty regarding animal tests. Avoiding all drugs during pregnancy, except in extremes, seems a drastic response. Studies of the effect of polymorphism in drug metabolism on teratogenic tests appears to us to be urgently needed. Comparisons of metabolism between different strains of rabbits, beyond the present selection of strains for thalidomide sensitivity, are needed.

Species variations in the N-methylation of pyridine have been reported by D'Souza, et al. (31). Cats, gerbils, guinea pigs, and hamsters are EM, while humans, mice, rabbits, and rats are PM (Table I). The mouse, rabbit, and rat are, thus, good surrogates for man for amines. Since methylation to quaternary amines could represent an intoxication step, experiments with EM would be more conservative.

All of these studies point to the need for a greater precision in examining drug metabolism in animals, with regard not only to the species chosen, but also to the strain chosen. Strains mimicking one or more human phenotype should be included in each compound evaluation.

Table I. Species Variations in N-Methylation of Pyridine

Species	Total Excreted	% Dose Excreted in 24 hrs. N-Methylpyrridinium Excreted
Extensive Methylators		
Cat	75	40
Gerbil	52	26
Guinea Pig	66	30
Hamster	67	26
Poor Methylators		
Man	67	9
Mouse	66	12
Rabbit	51	19
Rat	48	5

Conclusion

Animals continue to be fair surrogates for man, despite marked differences. Anatomical variations are important, since they can alter the quantitative response of test animals. The upper respiratory tract is particularly relevant in this regard for inhalation exposures of animals. Particles inhaled by man may be excluded from the lower respiratory tract of rodents, because of the smaller diameter of the airway and the greater filtration of particles in the nasopharyngeal cavity. While the lower respiratory tracts of rodents and man also differ, quantitative morphometric studies have improved maps of this area to the point at which they are useful in mathematical modeling. Using the physiological-anatomical approach to kinetic modeling, accurate predictions can be made for drug toxicity in man based on animal studies. Hopefully, the inhalation modeling of aerosols and gases will be validated shortly and will add this dimension to prediction of human toxicity from exposure to these toxic atmospheres.

Polymorphism in oxidative metabolism by man adds significant complexity to drug and toxicant testing. If oxidative metabolism of xenobiotic compounds continues to be considered a major determinant in toxicity, carcinogenicity, and teratogenicity, then animal surrogates will have to be chosen with the characteristics of drug metabolism in mind. The lack of oxidative metabolism in man is associated with adverse drug reactions due to higher blood levels of drugs; e.g. greater apparent potency. The lack of such metabolism in animals results in false negative errors for tests in which the metabolite is the ultimate toxicant; e.g. selectivity in teratogenicity in rodents. Polymorphism in drug metabolism is presumably due to genetic control over the induction and type of P-450 isoenzyme present in the tissues. Not only are fewer metabolites formed by PM, but the products are different. Some minor metabolites may be more toxic than the major ones. The matter is complex and not amenable to intuitive analysis. One could argue that rapid metabolism leads to rapid elimination, but rapid metabolism could lead to higher local concentrations of reactive metabolites and toxicity by overcoming detoxification pathways. Slower metabolism could lead to larger amounts of unreacted drug and, therefore, to longer exposure to both parent drug and its metabolites. If the parent compound is a drug or toxicant in its own right, PM leads to greater toxicity. PM could also lead to longer exposure to low levels of reactive metabolite, which in turn could lead to greater toxicity. A quantitative analysis using kinetic modeling appeals to us as a solution to this dilemma. Obviously, much greater comparative detail is needed to assure the continued usefulness of animal surrogates in predicting human toxicity.

Literature Cited

1. Weibel, E. R. Morphometry of the Human Lung. Academic Press: New York, 1963, 151 pp.
2. Klimet, V. Folia Morphol. 1973, 21, 59.
3. Schreider, J. P.; Raabe, O. G. Anat. Rec. 1981, 200, 195.
4. Miller, F. J.; Menzel, D. B.; Coffin, D. L. Environ. Res. 1978, 17, 84.
5. Lamb, D.; Reid L. J. Pathol. Bacteriol. 1968, 96, 97.
6. Reid, L. Arch. Intern. Med. 1970, 126, 428.
7. Reid, L. M.; Jones, R. Environ. Health Persp. 1980, 35, 113.
8. Himmelstein, K. J.; Lutz, R. J. J. Pharmacol. Biopharmac. 1979, 7, 127.
9. Lutz, R. J.; Dedrick, R. L.; Matthews, H. B.; Eling, T. E.; Anderson, M. W. Drug Metabol. Disp. 1977, 5, 386.
10. Lutz, R. J.; Dedrick, R. L.; Zaharko, D. S. Pharmac. Ther. 1980, 11, 559.
11. Bischoff, K. B.; Dedrick, R. L.; Zaharko, D. S.; Longstreth, J. A. J. Pharm. Sci. 1971, 60, 1128.
12. Bischoff. K. B. Cancer Chemother. Reports. Part 1. 1975, 59, 777.
13. Freeman, G.; Juhos, L. T.; Furiosi. N. J.; Mussenden, R.; Stephens, R. J.; Evans, M. J. Arch. Environ. Health. 1974, 29, 203.
14. Astarita, G. Mass Transfer with Chemical Reaction; Elsevier: New York, 1967, p. 53.
15. Stephens, R. J.; Sloan, M. F.; Evans, M. J.; Freeman, G. Amer. J. Pathol. 1973, 74, 31.
16. Stephens, R. J.; Sloan, M. F.; Evans, M. J.; Freeman, G. Exp. Mol. Pathol. 1974, 20, 11.
17. Miller, J. A.; Miller. E. C. in Origins of Human Cancer; Hiatt, H. H.; Watson, J. D.; Winsten, J. A. Eds: Cold Spring Harbor Laboratory: Cold Spring Harbor, New York, 1977, p. 605.
18. Price-Evans, D. A.; Mahgoub, A.; Sloan, T. P.; Idle, J. R.; Smith, R. L. J. Med. Genet. 1980, 17, 102.
19. Mahgoub, A.; Idle, J. R.; Dring, L. G.; Lancaster. R.; Smith, R. L. Lancet 1977, 2, 584.
20. Mahgoub, A.; Idle, J. R.; Smith, R. L. Xenobiotica 1979, 9, 51.
21. Sloan, T. P.; Mahgoub, A.; Lancaster, R.; Idle, J. R.; Smith, R. L. Br. Med. J. 1978, 2, 655.
22. Idle, J. R.; Sloan, T. P.; Smith, R. L.; Wakile, L. A. Br. J. Pharmacol. 1979, 66, 430.
23. Idle, J. R.; Ritchie, J. C.; Smith, R. L. Br. J. Pharmacol. 1979, 66, 432.
24. Kitchen, I.; Tremblay, J.; Andre, J.; Dring, L. G.; Idle, J. R.; Smith, R. L.; Williams, R. T. Xenobiotica 1979, 9, 397.

25. Danhof, M.; Idle, J. R.; Teunissen, M. W. E.; Sloan, T. P.; Breimer, D. D.; Smith, R. L. Pharmacology 1981, 22, 349.
26. Al-Dabbagh, S. G.; Idle, J. R.; Smith, R. L. J. Pharm. Pharmacol. 1981, 33, 161.
27. Ritchie, J. C.; Sloan, T. P.; Idle, J. R.; Smith, R. L. Ciba Foundation Symposium 1980, 76, pp. 219.
28. Speilberg, S. P. NEJM 1982, 307, 115.
29. Phelan, M. C.; Pellock, J. M.; Nance, W. E. NEJM 1982, 307, 99.
30. Shum, S.; Jensen, N. M.; Nebert, D. W. Teratology 1979, 20, 365.
31. D'Souza, J.; Caldwell, J.; Smith, R. L. Xenobiotica 1980, 10, 151.

RECEIVED November 4, 1983

4

Basic Concepts of the Dose-Response Relationship

ROBERT SNYDER

Joint Graduate Training Program in Toxicology, Rutgers, The State University of New Jersey, and College of Medicine and Dentistry of New Jersey, Piscataway, NJ 08854

The dose-response relationship is the cornerstone of Pharmacology/Toxicology. It quantitatively defines the role of the dose of a chemical in evoking a biological response. In the absence of chemical no response is seen. As chemical is introduced into the system the response is initiated at the threshold dose and increases in intensity as the dose is raised. Ultimately a dose is reached beyond which no further increase in response is observed. The dose-response relationship can be demonstrated for interactions of chemicals with biological receptors leading to physiological responses, therapeutic effects of drugs, or for toxic, lethal, teratogenic, mutagenic or carcinogenic effects of chemicals. The data from these studies can be expressed as dose-response curves which can take the form of linear plots or a variety of reciprocal or logarithmic transformations.

Two types of dose-response relationships are observed. The first is the incremental change in response of a single system or individual as the dose is increased. The second is the distribution of reponses in a population of individuals given different doses of the agent. The former are frequently used for the determination of the mechanism of interaction between the chemical and the biological system. The latter describe the response of a population of individuals and can also be used to determine multimodal responses indicative of genetic variations.

The dose-response relationship is of key importance when attempting to define allowable exposure of humans to chemicals in the workplace, consumer products or the environment. Usually initial studies are done in animals and, where

0097-6156/84/0239-0037$06.00/0

possible, they are compared with data derived from recorded human exposure. The reliability of extrapolations from these data is compromised by the inherent inaccuracy of the data observed in the high and, more importantly, the low dose regions of the dose-response curves since these usually demonstrate the fewest reponses. It is essential that we develop new approaches to understanding responses to low doses of chemicals if we are to define safe limits of exposure with accuracy.

The early history of Pharmacology and Toxicology was characterized by exploration of qualitative descriptions of the actions of drugs and toxic agents. Eventually a more quantitative approach had to be taken to pave the way for mechanistic studies. The necessity for quantitation of biological data was argued by A.J. Clark (1) who attempted to characterize cells as physico-chemical systems. He discussed the dose-response relationship in terms of controlling factors such as equilibria and kinetics in cell-drug interactions, and intracellular binding of drugs. It is clear that interactions between chemicals and biological systems demonstrate similarities regardless of the chemical studied. The first necessity is a chemical to be studied; the second is a biological assay system in which to study the chemical. In the absence of the chemical no response is observed. Upon addition of the chemical at a critical dose or concentration the response begins to be observed and this is called the "threshold." As the dose increases the response increases, however, the quantitative relationship between the increased dose and increased response may vary among chemicals and systems. Eventually the dose reaches a magnitude beyond which no further increment in response is seen. Beyond that dose only the maximum activity is observed. At extremely high doses for the responses being observed, the response is either lost or cannot be seen because a toxic effect of the chemical may come into play. However, over a reasonable concentration range the dose-response relationship is maintained.

Modern graphical analyses of dose-response phenomena are largely derived from the pioneering efforts of Trevan (2), Bliss (3) and Gaddum (4). This description, which makes liberal use of descriptive material compiled by Goldstein *et al.* (5) and Hayes (6), will investigate the modes of expression of dose-response curves making use of a variety of data transformations. Both incremental and quantal responses will be discussed. The application of these concepts to lethality, toxicity, carcinogenesis, teratogenesis and mutagenesis will be described.

Finally problems of the dose-response relationship relative to low dose exposure will be explored.

Bioassay Systems

The single most important entity in the study of the dose-response relationship is the bioassay system in which the chemical will be studied. Since the most essential feature of the results will be the quantitative data which are derived, the rules governing the accuracy and precision of the assay should approach as nearly as possible those achieved in measurements in chemical systems. Since biological systems are not machines, accuracy and precision can be difficult problems in bioassays. However, biological systems frequently are the match of chemical systems when it comes to sensitivity since the dose or concentration of chemical to which the bioassay systems may respond is often exceedingly low.

In chemical analyses the limits of accuracy relate to the relationship between the value observed and the actual value. The limiting feature is the method or the instrument used for the measurement. Since the actual value is often not known in an experimental situation, the determination will be based on the result of multiple measurements. If the differences between the results obtained in repeated determinations is small the measurement can be considered to be precise, i.e. reproducible. The limits to accuracy and precision in biological systems can be explored using three levels of biological organization as examples: whole animals, isolated organ systems, and purified enzymes.

Whole animals are used in many bioassay systems. The start of most safety evaluation studies involves determining the median lethal dose of the chemical, i.e. the LD_{50}. Since many animals are necessary for these studies small, relatively inexpensive rodents are usually used, e.g. mice or rats. Furthermore, outbred, i.e. genetically heterogeneous animals of the same strain, rather than the more exotic inbred, strains are used. This not only reduces the cost but avoids cases of genetically determined unusual sensitivity or resistance to the chemical. To be sure, the major problem in these studies is the assumption that one can extrapolate from the sensitivity of animals to the sensitivity of humans. While examples can be cited for unexpected differences in sensitivity between humans and specific animal strains to the lethality of a chemical, for the most part comparative lethality in animal strains to various chemicals is similar to the relative sensitivity of humans to the various chemicals. Thus, to use an extreme example, in rats and mice as well as in humans, sucrose is less toxic than cyanide. That does not mean that the LD_{50} for any given chemical is the same in all species. It is fortunate, however, that except for unusual examples, toxicity classes, i.e. ranges of doses in which chemicals are lethal, do not vary widely among species.

The accuracy of LD_{50} determinations cannot be verified in a given experiment since it can only be done once with one group of animals. Provided that normal healthy animals are used and the correct doses are administered by the proper route, the result must be accepted. The precision is another matter. Repetition of the study with animals of the same strain, sex, age, etc., may lead to somewhat different values because of biological variability. This can be dealt with by expressing the results in terms of confidence limits derived from a statistical evaluation of the data. These differences between experiments may not be great but it would not be unexpected if they were greater than those observed in chemical determinations. As a practical matter they are usually sufficiently accurate and precise for their intended purpose which is to indicate the relative lethality of the compound.

The more difficult problem with whole animals concerns events which occur over long periods of time. The LD_{50} value must always be accompanied with an indication of the time over which the animals were observed before the experiment is terminated. If not, every treatment would be considered lethal since every animal dies eventually, or no chemical would be considered lethal since both control and treated animals would die eventually. Thus, observation periods of 24 hours or two weeks are often chosen as end points. When dealing with carcinogenesis, however, the time of the study is considered the life time of the animal which in the case of mice or rats may extend to two years or more. Furthermore, since control animals may display spontaneous tumors and the tumor incidence in both treated and control animals may be small, the total number of animals in the experiment often plays a key role in determining the accuracy of the results. The responses discussed here are classified as quantal since each animal provides only one piece of data. The animal either dies or it does not; it develops tumors or it does not. The same observation cannot be repeated in the same animal and the effect of a higher dose in that animal cannot be investigated.

In contrast a number of isolated organ preparations have been used as bioassay systems. Historically bioassay systems were developed when the nature of the chemicals themselves were often unknown and/or the sensitivity of chemical methods was insufficient to measure the extremely small concentrations of chemicals necessary to produce responses in bioassay systems. Thus, these systems could be used not only to measure the effect of the chemical on the system, but once the system was calibrated the concentration of a solution of the chemical could be determined based on the response it produced in the system. Furthermore, bioassay systems allowed for the demonstration of specific principles. For example, the demonstration by Loewi

(7) that a chemical mediator controlled heart rate depended upon the demonstration that blood flowing from one frog heart contained a substance which could slow down the rate of a second heart. This was a bioassay system in which two isolated frog hearts were used. Other systems such as the response of muscle preparations in tissue baths to direct stimulants of contraction, such as the isolated clam heart, or the isolated cat spleen are based on the ability to measure changes in contraction of the organs in reponse to chemicals. The latter will be used for some of the examples cited below. The feature which distinguishes these systems from the whole animal systems described above is that the responses which can be measured are incremental. Thus, the addition of a given concentration can produce a response of a given magnitude but the same preparations can then be treated with a higher concentration and a greater response observed. Clearly, there are advantages to observing changes in response in the same system to either different chemicals or the same chemical at different doses. Qualitative differences and similarities are emphasized and quantitative differences can be be evaluated with greater certainty.

Finally, isolated enzymes, which come closest to working with pure chemicals can be used to study the mechanisms of the effects of chemicals. The interaction of chemicals with biological receptors follow much the same laws as the interaction of substrates with enzymes. Thus, parallels can be drawn between the interaction of chemicals with receptors and mechanisms of enzyme catalyzed reactions. The main difference is that receptors dissociate from chemicals leaving the chemicals unchanged whereas enzymes alter the chemicals.

Graphical Presentation of the Dose-Response Relationship

The dose-response relationship can be expressed graphically using a variety of mathematical transformations. In the simplest expression the dose is plotted on the abcissa and the response on the ordinate. Both are expressed in appropriate units on an arithmetic basis (Figure 1). Although the data are expressed without further transformation the result is not a straight line throughout. The initial slope tends to be straight and is often the section of the curve which is of greatest interest. Thus, Figure 2 shows the straight lines obtained expressing an increase in mutagenesis when either strains TA 1535 or TA 100 of *Salmonella typhimurium* are exposed to increasing concentrations of sodium azide (8).

Figure 1 is typical of an incremental dose-response curve observed using a preparation in which a muscle is fixed in a bath with one end tied to a device for recording changes in tension and the dose of chemical agent, i.e. an agonist, which modifies tension is varied. If it is assumed that (1) the

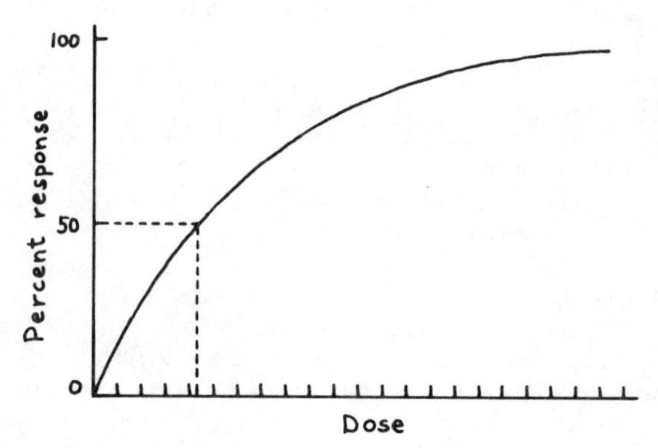

Figure 1. The relationship between dose and response plotted arithmetically.

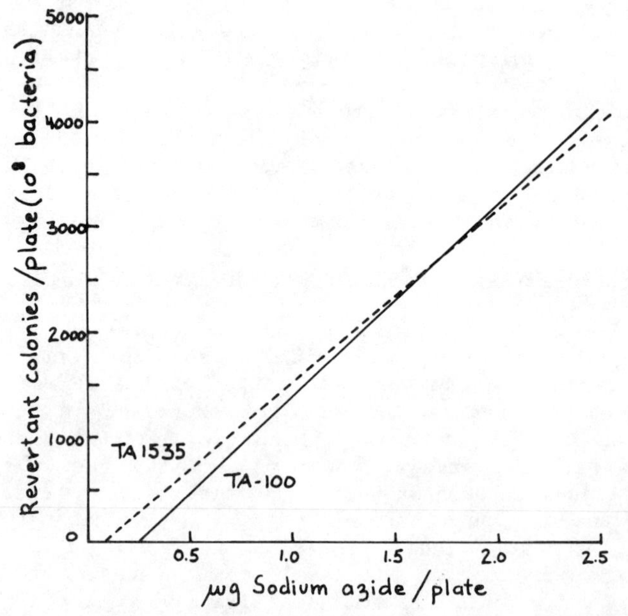

Figure 2. Dose-response curve for mutagenicity of sodium azide in two strains of Salmonella typhimurium plotted arithmetically.

response is proportional to the number of receptors occupied by the agonist, (2) one mole of agonist reacts per receptor site, and (3) the total number of receptors is much smaller than the number of agonist molecules, an equation can be derived which describes the interaction of agonist and receptor. It describes the interaction in terms of the number of receptors occupied. Thus, in the absence of agonist no receptors are occupied. When all receptors are occupied the maximum response is observed. At increasing doses in between incremental responses can be observed. In the course of the interaction the agonist reacts with the receptor and then dissociates allowing another agonist molecule to approach the receptor.

Thus, if A=agonist, R=receptor, k_1 and k_2 are rate constants and K_A is the dissociation constant, the following equations can be written:

$$[R] + [A] \underset{k_2}{\overset{k_1}{\rightleftharpoons}} [RA] \quad (1)$$

If rsp= response and Rsp_{max}= maximum response the following equation can be derived:

$$rsp = \frac{Rsp_{max}\ A}{K_A + A} \quad (2)$$

This is the equation for the curve seen in Figure 1. It is in most respects identical to the Michaelis-Menton equation:

$$v = \frac{V_{max}\ S}{K_M + S} \quad (3)$$

The only difference is that in enzymatic reactions described by the Michaelis-Menton equation substrate is consumed and, therefore, K_M is not a true dissociation constant whereas in equation (2) K_A is a true dissociation constant.

In Figure 1 the dissociation constant can be obtained by determining the dose of agonist necessary to give half of the maximal response. Because we are dealing with a curve, however, it is difficult to determine this value with accuracy from the arithmetic dose-response plot. The data can be expressed as a straight line most readily by applying the technique of Lineweaver and Burk (9) and plotting the data as the reciprocal of both dose and response (Figure 3). The equation describing the resulting straight line is:

$$\frac{1}{rsp} = \frac{K_A}{Rsp_{max}} \cdot \frac{1}{A} + \frac{1}{Rsp_{max}}$$

The maximum response can be derived from the point on the

ordinate that intersects the straight line. The dose giving half of the maximum response can then be easily derived and is the dissociation constant.

The double reciprocal plot has been used extensively in the study of enzymatic reactions to characterize the rate of the reaction, the Michaelis constant, and the mode of action of inhibitors. It can also be used to study the interaction of chemicals with biological systems. The simplest types of interactions can be illustrated in Figures 4 and 5. The lowest lines in each represent the dose-response relationship for a hypothetical system. When the action of the agonist is inhibited by another chemical, i.e. an antagonist, the response is reduced and two upper lines represent the degree of antagonism as a function of dose of the agonist, each line representing a different dose of antagonist. In Figure 4 all three lines intersect at the ordinate. These data are interpreted to mean that the agonist and antagonist are probably reacting at the same site. The reaction of each with the receptor site is reversible because by increasing the dose of the agonist it is possible to completely overcome the effects of the antagonist. Thus, the maximum response is not altered. This is called competitive antagonism since the two agents compete for the same receptor site. The dissociation constant can be calculated for the agonist-receptor interaction from the point where the straight line obtained in the absence of antagonist crosses the abcissa. In contrast Figure 5 demonstrates the double reciprocal plot characteristic of non-competitive antagonism. Note that the three lines intercept at the abcissa rather than at the ordinate at a point which is the negative reciprocal of the dissociation constant. On the ordinate the maximum response in the presence of antagonist is in each case smaller than that produced by the agonist alone. Thus, regardless of the size of the dose of agonist the effects of the antagonist cannot be completely overcome. Mechanistically this suggests that either the antagonist reacts at a site remote from the site at which the agonist acts or the antagonist reacts irrereversibly with the receptor and thereby decreases the total number of active receptor sites.

A specific example of a competitive antagonist in a figure taken from a paper by Chen and Russell (10) can be seen in the effect of diphenhydramine, an anti-histamine on the histamine-induced decrease in blood pressure in the dog (Figure 6). Note that with increasing dose of diphenhydramine the effect of histamine is decreased but by increasing the dose of histamine the antagonistic effects are eventually overcome. In contrast they showed that when ergotamine, a vasoconstrictor, which raises blood pressure by a mechanism remote from the effect of histamine, is adminstered with histamine, the antagonism cannot be completely overcome by increasing the dose. This type of antagonism is not competitive.

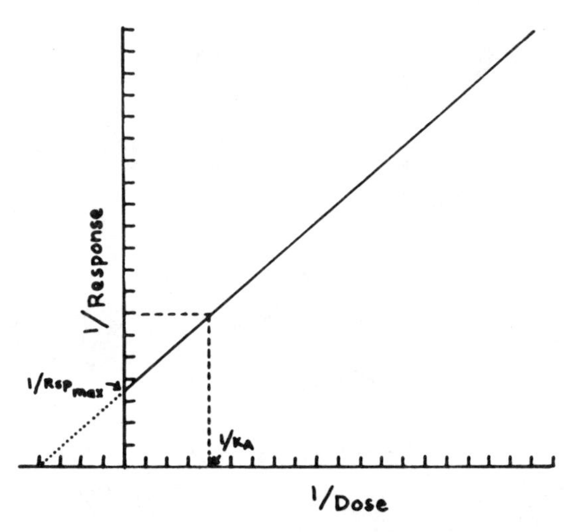

Figure 3. Schematic presentation of double reciprocal dose-response plot.

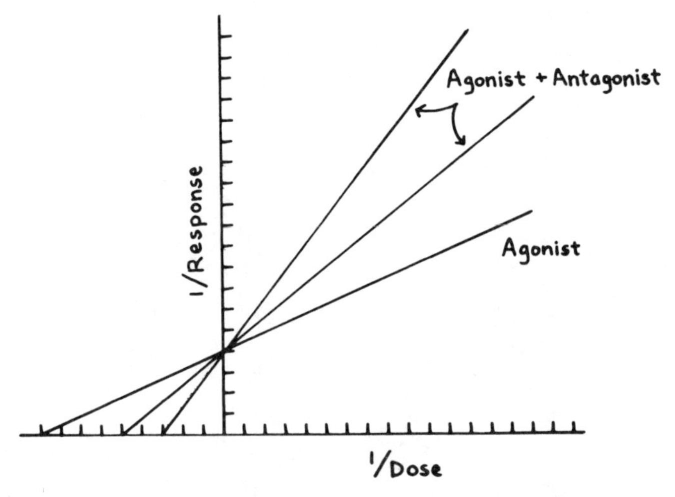

Figure 4. Schematic presentation of competitive antagonism using a double reciprocal plot.

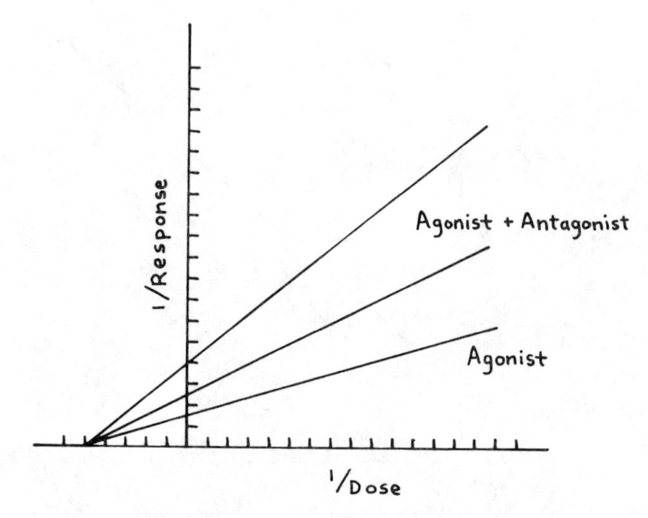

Figure 5. Schematic presentation of non-competitive antagonism using a double reciprocal plot.

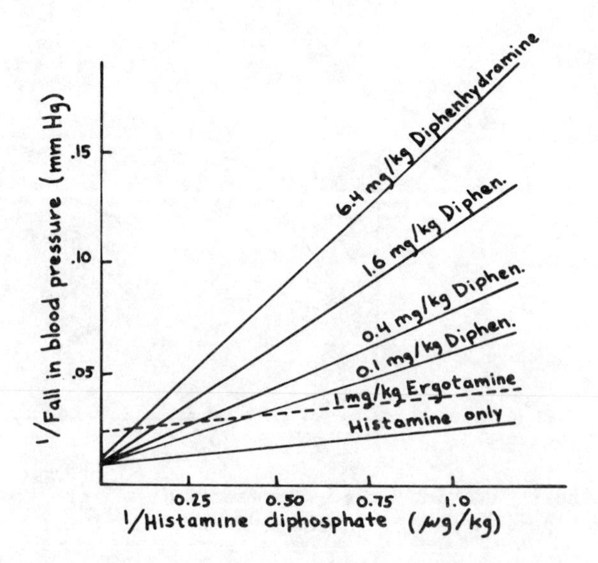

Figure 6. Double reciprocal plot demonstrating antagonism of diphenyhydramine and ergotamine to the blood pressure lowering effects of histamine in the dog. (Reprinted with permission from Ref. 10.)

In addition to arithmetic representations of the data and reciprocal transformation, it is common to plot biological data using logarithmic transformations. Expression of the dose in logarithmic terms allows for the description of the effects over a wide range of doses on a simple scale. An idealized semi-log-dose-response curve is shown in Figure 7. The ordinate represents the percent of maximum response attainable in the bioassay system, which for the purposes of this discussion produces an incremental response, and the abcissa is the dose plotted in logarithmic units over a range covering three orders of magnitude. The dose producing half of the maximal response is called the ED_{50}, i.e. the dose giving 50% of the maximal response. An example of the practical application of this technique is taken from a paper by Bickerton (11) who investigated the effects of catecholamines on the isolated cat spleen. The spleen contracts when stimulated by epinepherine (epi) or norepinepherine (n-epi). The degree of contraction can be measured on a strain gauge and increases as the concentration of catecholamine in the system is raised. Figure 8 shows a dose-response curve comparing the relative activity of epi and n-epi in this system. The dose expressed logarithmically covers more than a 10,000 fold concentration range. Both appear to produce the same maximum response, i.e. both have the same efficacy. The effects of epi appear to be produced at lower doses and, thus, for this system, epi is said to be more potent than n-epi. The shape of the curve, i.e. "S" shaped, is characteristic of these transformations. Generally speaking for curves of this type the middle portion of the curve tends to approximate a straight line. The slope of the curved is determined by the dosage range required to observe the entire dose-response relationship.

Anatagonism can be explored using semi-log transformations. Thus Bickerton (11) examined the effects of two types of antagonists on the the effects of n-epi on the cat spleen. Figure 9 shows the log dose-response curve for n-epi at the left and the dose response curves obtained with the same doses of n-epi when tolazoline was added at either of two concentrations at the right. The effects of n-epi can still be observed but higher doses of n-epi were required to produce the same effect. When the dose of n-epi was raised sufficiently high the effect of tolazoline was completely overcome. Thus, tolazoline is a competitive antagonist of n-epi. In contrast, Figure 10 shows the effect of addition of dibenamine at either of two concentrations. Again n-epi is less potent in the presence of the antagonist, but in addition it is not possible to overcome the effects of dibenamine regardless of how high the dose of nor-epi is made. Dibenamine is a non-competitive antagonist and it is known that it binds irreversibly to receptors, thereby causing inactivation and hence reducing the total number of receptors available for stimulation by n-epi.

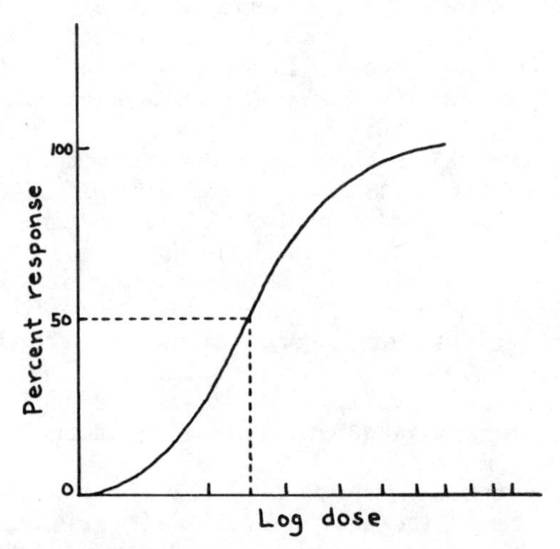

Figure 7. Schematic presentation of a log dose-response curve.

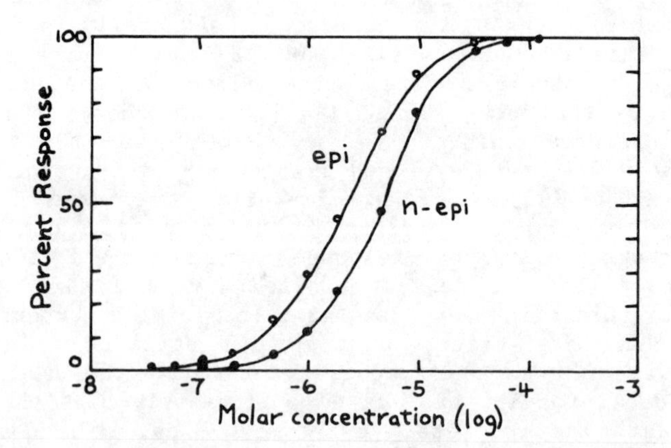

Figure 8. Log dose-response curve for the effects of epinepherine and norepinepherine on the isolated cat spleen bioassay system. (Reprinted with permission from Ref. 11.)

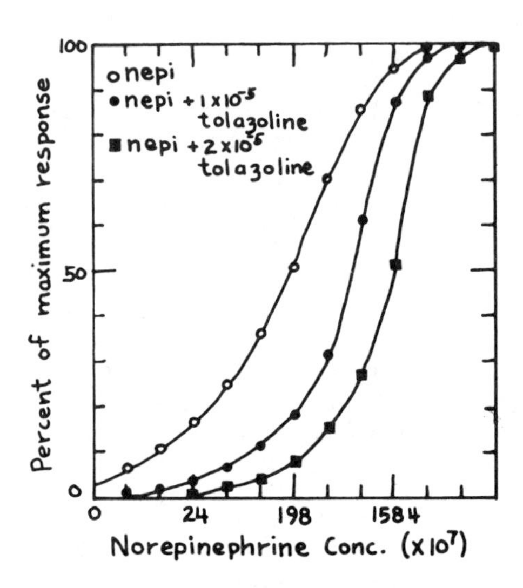

Figure 9. Demonstration of competitive antagonism of the of the effect of norepinepherine on the isolated cat spleen preparation by tolazoline. (Reproduced with permission from Ref. 11.)

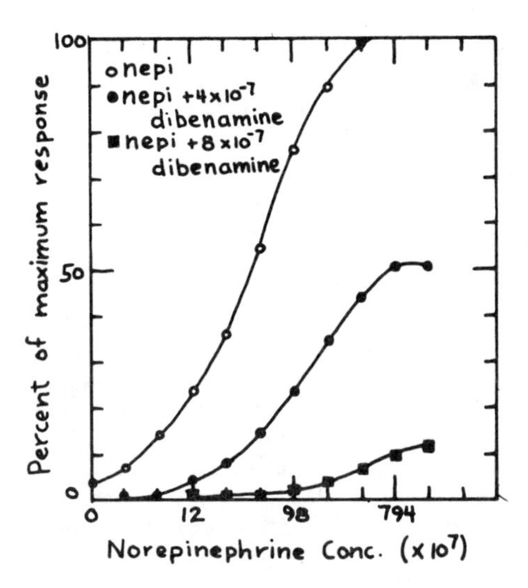

Figure 10. Demonstration of the noncompetitive antagonism of the effect of norepinepherine on the isolated cat spleen preparation by dibenamine. (Reproduced with permission from Ref. 11.)

Although the "S" shape of the semi-log plot represents a reasonable illustration of the dose-response relationship, log-log transformations are often preferred because they yield straight lines. Thus, Harbison and Becker (12) investigating teratogenic effects of diphenylhydantoin in mice compared the dose response relationship for the production of orofacial versus skeletal abnormalities using a log-log plot (Figure 11). In this case the concern was not the wide range of doses but the examination of the steep slopes which suggested that the dose range over which the abnormalities were observed was quite narrow.

The log-log transformation has been used extensively in evaluating lethality and carcinogenesis in populations of humans or animals. These are treated as quantal responses and the concern is whether or not a response occurred rather than the magnitude of the response, i.e. the animal died or it did not; the animal developed tumors or it did not. Figure 12 shows an example of a hypothetical semi-log dose-response curve indicative of the accumulated quantal responses in a population as the dose is raised. Superimposed on the "S" shaped curve is the same data plotted as a frequency distribution in which the ordinate represents the increment in responses as the dose is elevated. The frequency distribution approaches a normal Gaussian curve. Thus, at low doses few members of the population respond. With increases in dose a greater percentage of the population responds. At the ED_{50} half of the population has responded and half has not. As the dose increases additional individuals respond but the curve slopes down because those sensitive only at the higher doses represent an increasingly smaller segment of the population. Eventually at the highest doses the most resistant individuals eventually respond. Since this is a normal distribution the abcissa can be expressed as the dose itself or as the median dose $\pm$one or more standard deviations.

In addition to expressing the dose as a function of the median and standard deviations, it is also possible to express the response in the same way. For this purpose the concept of the normal equivalent deviation (NED) i.e., the number of standard deviations on either side of the median response, has been devised and can be used as a means of expressing the response. To avoid the use of positive and negative numbers, and recognizing that it is likely that data will not frequently be collected which lies more than a few standard deviations from the mean, a convention has been adopted called the probit. The number 5 is added to the NED to yield positive numbers and indicates the number of standard deviations from the mean that the response is found. Table I shows the relationship between percentage response, NED and the probit value.

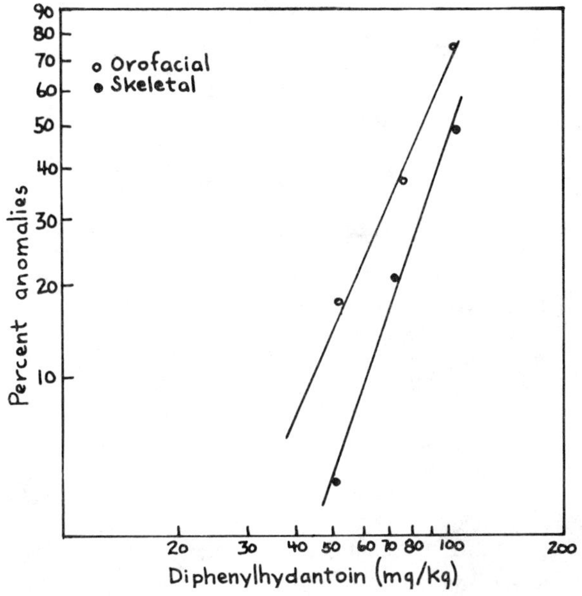

Figure 11. Use of the log dose-response plot to investigate the teratological response of mice to diphenylhydantoin. (Reproduced with permission from Ref. 12.)

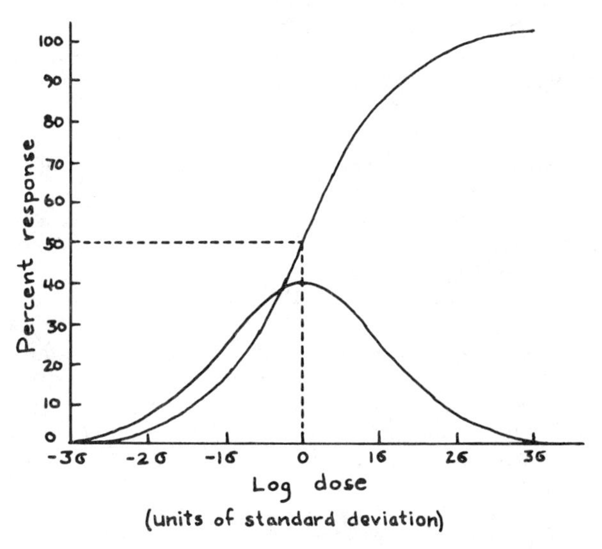

Figure 12. Schematic plot of log dose-response replotted as a frequency distribution.

TABLE I

% Responding	NED	Probit
2	-2	3
16	-1	4
50	0	5
84	+1	6
98	+2	7

Figure 13 shows a log-probit plot examining the lethality of three chemicals. Note that all are straight lines and the point at which they pass through probit 5 can be extrapolated to the abscissa to obtain the LD_{50}, i.e. the dose which would be expected to be lethal to half of a given population. Lines A and B are parallel but A is more potent than B. C shares the same LD_{50} with B but the slope is steeper. Thus, the dose range over which lethality is observed with B is greater than C. Using this approach it is also possible to predict doses which would be lethal to 25% of the population, 10% etc.

The most difficult problem is encountered when attempting to determine responses by a small segment of the population, i.e. less than 10%. It is most frequently encountered when attempting to extrapolate the likelihood of cancer in a population exposed to low doses of carcinogen. Because once initiated carcinogenesis is largely an irreversible process and many authorities believe that a single interaction of carcinogen with DNA is sufficient to initiate carcinogenesis, they argue that the lower end of the probit plot is linear down to the nearest possible approximation of zero dose. Unfortunately, as the dose is lowered the numbers of individuals responding decreases and the problem of spontaneous tumors also begins to interfere. Figure 14 shows an example of the problem taken from Bryan and Shimkin (13). The data reflects 3-methylcholanthrene-induced carcinogenesis and clearly at the higher dose ranges a straight line is attained with some certainty since a relatively large number of responses is observed. The authors question whether data obtained at the lower end of the plot is also linear. Some authorities argue that the shape may change because of differences in pharmacokinetics and xenobiotic metabolism at low doses but this remains a matter of debate. A variety of alternative extrapolation techniques have been developed (14) and considerable effort is underway to attempt to fit real data to the theorized approaches.

Conclusion

The aim of this discussion was to characterize the dose-response relationship and the approaches used in its study. Early

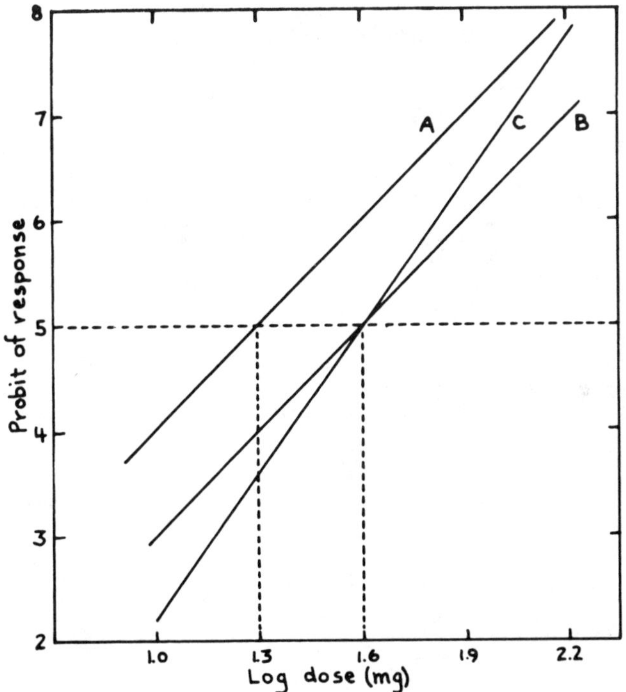

Figure 13. Schematic demonstration of a log dose-probit plot.

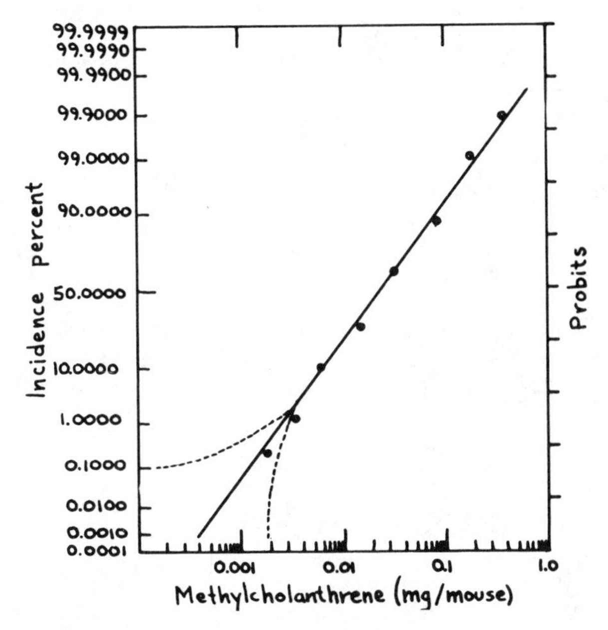

Figure 14. Log dose-response plot for 3-methylcholanthrene induced carcinogenesis. (Reproduced with permission from Ref. 13.)

developments in drug-receptor interactions built up a literature based on incremental responses within defined systems. Using this approach attempts to extrapolate to the quantal responses seen in human populations have raised controversy pimarily in the area of low dose, long term exposure where dose-response curves may not be readily predicted. Public health authorities, in attempting to protect the public from exposure to potential carcinogens have taken the approach that a straight line extrapolation in which no tumors would be expected only at zero dose is the most conservative attitude and in the best interests of the public. Regardless of these considerations it remains the responsibility of the toxicologist to define the shape of these dose response curves based on theoretical and empirical studies. In the area of carcinogenesis these studies will include examples of both pharmacokinetics and molecular biology and the problems may not be solved soon. Nevertheless, it is essential that work in this area continue with vigor, not only to serve the needs of the regulators but to provide a scientific basis for understanding the etiology of these diseases in the population.

Acknowledgment

The author wishes to thank Mr. James Griffiths for the preparation of the figures.

Literature Cited

1. Clark, A.J. "The Mode of Action of Drugs on Cells"; Edward Arnold and Co., London, 1933.
2. Trevan, J.W., The error of determination of toxicity, Proc. Roy. Soc. Lond. (Biol.) 1927,101,483-514.
3. Bliss, C.I., The method of probits,Science, 1934, 79, 38-39.
4. Gaddum, J.H., Lognormal distributions, Nature, 1945,156, 463-466.
5. Goldstein, A.; Aronow, L; Kalman, S.M. "Principles of Drug Action"; Harper and Row, Publishers, New York, 1969.
6. Hayes, W. J., Jr. "Toxicology of Pesticides"; Williams and Wilkins, Baltimore, 1975.
7. Loewi, O. Ueber humorale Uebertragkeit der Herznervenwirkung, Arch. f. ges. Physiol. 1921, 189, 239-242.
8. Witmer, C.; Cooper, K.; Kelly, J. in Biological Reactive Intermediates-II B; Snyder, R., et al, Eds.; EFFECTS OF PLATING EFFICIENCY AND LOWERED CONCENTRATION OF SALTS ON MUTAGENICITY ASSAYS WITH AMES' SALMONELLA STRAINS Vol. 136 B, Plenum Publishers: New York, 1982; p. 1271-1284.
9. Lineweaver, H.; Burk, D. The determination of enzyme dissociation conastants, J. Am. Chem. Soc. 1934,56,658-666.

10. Chen, G.; Russell, D. A quantitative study of blood pressure response to cardiovascular drugs and their antagonists. J. Pharmacol. Exp. Therap. 1950, 99, 401-408.
11. Bickerton, R.K. The responses of isolated strips of cat spleen to sympathomimetic drugs and their antagonists. J. Pharmacol. Exp. Therap. 1963, 142,99-110.
12. Harbison, R.D.; Becker, B.A. The effect of phenobarbital and SKF 525A pretreatment on diphenylhydantoin teratogenicity in mice. J. Pharmacol. Exp. Therap. 1970, 175,283-288.
13. Bryan, W.R.; Shimkin, M.B. Quantitative analysis of dose-response data obtained with three carcinogenic hydrocarbons in strain C3H male mice. J. Natl. Cancer Inst. 1943, 3,503-531.
14. Food Safety Council, Scientific Committee, Quantitative risk assessment, in Proposed System for Food Safety Assessment 1978, Food and Cosmetics Toxicology,16, Supplement 2, 109-136.

RECEIVED November 4, 1983

5

High- to Low-Dose Extrapolation in Animals

CHARLES C. BROWN

National Cancer Institute, Bethesda, MD 20205

Quantitative risk assessment requires extrapolation from results of experimental assays conducted at high dose levels to predicted effects at lower dose levels which correspond to human exposures. The meaning of this high to low dose extrapolation within an animal species will be discussed, along with its inherent limitations. A number of commonly used mathematical models of dose-response necessary for this extrapolation, will be discussed. Other limitations in their ability to provide precise quantitative low dose risk estimates will also be discussed. These include: the existence of thresholds; incorporation of background, or spontaneous responses; modification of the dose-response by pharmacokinetic processes.

In recent years, as the serious long-range health hazards of environmental toxicants have become recognized, the need has arisen to quantitatively estimate the effects upon humans exposed to low levels of these toxic agents. Often inherent in this estimation procedure is the necessity to extrapolate evidence observed under one set of conditions in one population group or biological system to arrive at an estimate of the effects expected in the population of interest under another set of conditions.

The quantitative assessment of human health risk from exposure to toxic agents has been approached by relating the exposure level of the suspect to measures of health risk on the basis of either epidemiologic or clinical data on human populations or experimental data on animals or other biological systems. Unfortunately, there are often serious limitations with both approaches. Since human populations cannot be regarded as experimental subjects with regard to deleterious effects on health, the observational data from such sources are often incomplete and not of the desirable form and substance. Attendant with epidemiologic studies are difficulties in the accurate measurement of individual exposure patterns and the control of factors that may modify or confound the quantitative measures of health risk. Moreover, long delays often occur between exposure and the occurrence of a measureable effect. Such delays can range up to decades as seen in many cases of carcinogenesis associated with occupational exposure to certain agents, such as asbestos induced lung cancer.

Often by necessity, the potentially deleterious effects of chemical compounds must be tested in laboratory animals. For the extrapolation of animal study results to man, much care should be placed in the design and conduct of these studies, since many factors may influence their results. These factors include the dosage and frequency of exposure, route of administration, species, strain, sex and age of the animal, duration of the study, and various other modifying factors as deemed important for the particular agent and effect being studied.

Attendant with information on the dose-response of the agent in question is the necessity that the experimental data must be based on exposure levels higher than those for which the risk estimation is to be made. Some consideration has been given to the possibility of conducting extremely large experiments at very low dose levels. Use of large numbers of experimental subjects is necessary to reduce the statistical error so that very small effects can be adequately quantified. However, as Schneiderman, et al. (1) remark, "purely logistical problems might guarantee failure." Therefore, to obtain reliably measureable effects, the experimental information must be based on levels of exposure high enough to detect positive results. Since large segments of the human populations are often exposed to much lower levels, these high exposure level data must be extrapolated to lower levels of exposure. The purpose of this report is to describe the current statistical methods used for this "high to low dose" extrapolation in experimental animal species and to indicate the uncertainties necessarily attached to the estimates made with these methodologies.

The high to low dose extrapolation problem is conceptually straight-forward. The probability of a toxic response is modeled by a dose-response function P(D) which represents the probability of a toxic response when exposed to D units of the toxic agent. A general mathematical model is chosen to describe this functional relationship, its unknown parameters are estimated from the available data, and this estimated dose-response function P(D) is then used to either: (1) estimate the response measure at a particular low dose level of interest; or (2) estimate that dose level corresponding to a desired low level of response (this dose estimate is commonly known as the virtually safe dose, VSD).

Many mathematical models of this dose-response relationship have been proposed for this problem. The following section describes the models currently being used. One of the major difficulties inherent in this high to low dose extrapolation problem is that the estimates of risk at low doses, and correspondingly the estimates of VSD's for low response levels, are highly dependent upon the mathematical form assumed for the underlying dose-response. These difficulties are discussed in later sections.

Mathematical Models of Dose-Response

To estimate the effects expected to be observed outside the range of the experimental data, a mathematical model relating dose, i.e., level of exposure to the toxic agent, to response, i.e., a quantitative measure of the deleterious effect produced, is necessary. In general terms, dose-response is the relation between a measureable stimulus, physical, chemical or biological, and the response of living matter measured in terms of the reaction produced over some range of the degree or level of the stimulus.

The reactions to any one stimulus may be multiple in nature, e.g. loss of weight, decrease in organ function, or even death. Each reaction may have its own unique relation with the level of the stimulus. In addition, the measure of any specific reaction may be made in terms of the magnitude of the effect produced, quantitative response, whether or not a specific effect is produced, quantal response, or the time required to produce a specific effect, time to response. The discussion of models will be limited to quantal response models, but similar models may be used for responses measured in other units. These responses may be acute reactions, sometimes occurring within minutes of the stimulus, or they may be long-delayed effects such as cancer, which may not appear clinically until most of the subjects normal lifespan has elapsed. Other responses may not even appear in the exposed subject, but may become manifest in some later progeny.

The level of the stimulus, or dose level, may also be measured in different ways. For example, consider a subject that is exposed to a toxicant in its environment, either through the air breathed, the food eaten, or through some other external source of exposure. The dose level may be quantified in terms of concentration in the air or food, or in term of the quantity of the substance actually reaching the target receptor, some internal organ, or other tissue. The former may be thought of as the environmental, or "external", exposure level, while the latter may be termed the "internal" exposure level. Due to the subject's biochemical and physiological internal mechanisms, the dose-response may be quite different for the two measures of dose. Since the following material is applicable to dose as measured on any scale, no distinction between these two general bases of measurement will be made.

Tolerance Distribution Models

When the response is quantal, its occurrence for any particular subject will depend upon the level of the stimulus. For this subject under constant environmental conditions, a common assumption is that there is a certain dose level below which the particular subject will not respond in a specified manner, and above which the subject will respond with certainty. This level is referred to as the subject's tolerance. Because of biological variability among subjects in the population, their tolerance levels will also vary. For quantal responses, it is therefore natural to consider the frequency distribution of tolerances over the population studied. If D represents the level of a particular stimulus, or dose, then the frequency distribution of tolerances, f(D), may be mathematically expressed as

$$f(D)=dP(D)/dD$$

which represents the proportion of subjects whose tolerances lie between D and D+dD, where dD is small. If all subjects in the population are exposed to a dose of D_0, then all subjects with tolerances less than or equal to D_0 will respond, and the proportion, $P(D_0)$, this represents of the total population is given by

$$P(D_0) = \int_0^{D_0} f(D)dD$$

Assuming that all subjects in the population will respond to a sufficiently high dose level, then

$$P(\infty) = \int_0^{\infty} f(D)dD = 1$$

Figure 1 shows a hypothetical tolerance frequency distribution, f(D)dD, along with its corresponding cumulative distribution, P(D). Thus, when the response is quantal in nature, the function P(D) can be thought of as representing the dose-response either for the population as a whole, or for a randomly selected subject. The notion that a tolerance distribution, or dose-response function, could be determined solely from consideration of the statistical characteristics of a study population was introduced independently by Gaddum (2) and Bliss (3).

The results of toxicity tests have often shown that the proportion of responders increases monotonically with dose and exhibits a sigmoid relationship with the logarithm of the exposure level. This observation led to the development of the log normal, or probit, model for the tolerance frequency distribution,

$$f(D;\mu,\sigma) = (2\pi\sigma^2)^{-1/2} \exp - \frac{1}{2}\left(\frac{\log(D)-\mu}{\sigma}\right)^2 , \sigma>0$$

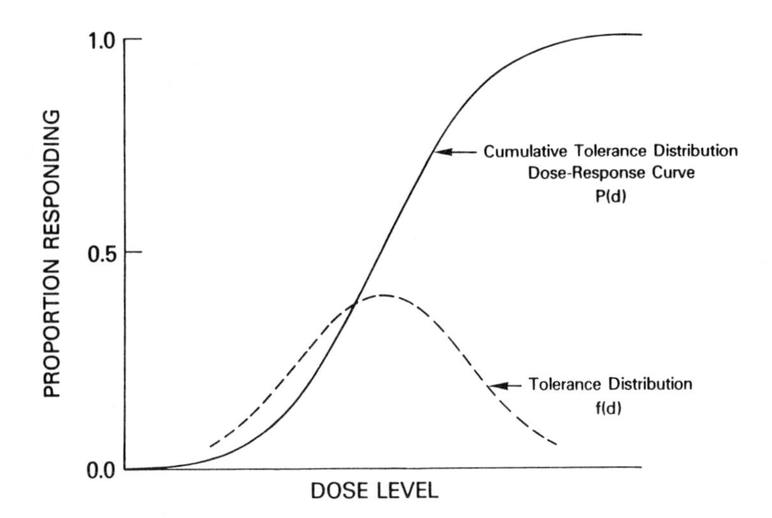

Figure 1. Relationship between tolerance distribution and dose-response curve.

while the dose-response function is given by the cumulative normal probability,

$$P(D;\mu,\sigma) = \Phi[(\log(D)-\mu)/\sigma]$$

where μ and σ^2 represent the mean and variance of the distribution of the log tolerances. This method was put into its modern form by Bliss (4), and Finney (5) gives a brief history of its development.

This dose-response model was originally proposed for use in problems of biological assay, i.e. the assessment of the potency of toxicants, drugs, and other biological stimuli, and has been primarily used for problems of dose-response interpolation (i.e. estimation within the range of observable response rates), rather than dose-response extrapolation (i.e. estimation outside the range of observable rates). Mantel and Bryan (6) and Mantel, et al. (7) later proposed its use, with suitable modification, for the problem of extrapolation of experimentally induced effects observed at "high" dose levels to those expected at "low" levels. Their modification was to assume a slope shallower than that observed in the experimental animal study. Their reasons for this modification were two-fold: (1) to conservatively guard against the possibility that the true dose-response in the "low" dose region might be different than that observed in the "high" dose region; and (2) inbred strains of laboratory animals are more likely to show steeper dose-response relationships than the heterogeneous human population to which the extrapolation is to apply. This assumed conservative slope is a key feature of the Mantel-Bryan methodology, though its choice is arbitrary. For the purpose of extrapolation, the particular slope selected is not meant to represent the "true" slope in the low dose region, but rather to represent a conservatively shallow slope no matter what the true dose-response may be in this region. Therefore, the Mantel-Bryan method was not proposed to provide necessarily valid estimates of low dose risk, but rather to provide "conservative" estimates of this risk. However, the "conservative" nature of this extrapolation methodology has been questioned by many authors (8-10).

Other mathematical models of tolerance distributions which produce a sigmoid appearance of their corresponding dose-response functions have been suggested. The most commonly used is the log logistic function,

$$P(D;a,b) = [1+\exp(a + b \log_{10}(D))]^{-1}, \quad b<0$$

which, like the log normal model is sigmoid and symmetric about the 50% response level, but approaches the extremes, 0% and 100% response, more slowly than does the log normal. The logistic function has been derived from chemical kinetic theory, and was proposed as dose-response model by Worcester and Wilson (11) and

Berkson (12). The log logistic and log normal functions are so similar in appearance that discrimination between them is nearly impossible. Other models have also been proposed, but do not have a wide acceptance and thus will not be discussed.

Models Derived From Mechanistic Assumptions

A number of dose-response models have been suggested on the basis of assumptions regarding the mechanism of action of the toxic agent upon its target site. The "hit theory" for interaction between radiation particles and susceptible biologic targets has generated a general class of these models (13). This theory is also applicable to the action of chemical toxicants upon their target sites. In general, this theory rests upon a number of postulates, which include: (1) the organism has some number M of "critical targets" (usually assumed to be infinitely large); (2) the organism responds if m or more of these critical targets are "destroyed"; (3) a critical target is destroyed if it is "hit" by k or more toxic particles; and (4) the probability of a hit in the low dose region is proportional to the dose level of the toxic agent, i.e. Prob(hit) = λD, $\lambda > 0$.

Some commonly used special cases of this general theory are the single-hit model,

$$P(D;\lambda) = 1-\exp(-\lambda D)$$

where the subject responds if a single critical target is destroyed by a single hit; and the multihit model,

$$P(D;\lambda,k) = \int_0^{\lambda D} \frac{x^{k-1}\exp(-x)}{\Gamma(k)} dx$$

where $\Gamma(k)$ denotes the gamma function, and the subject responds if a single critical target is destroyed by k hits. This multihit model, also referred to as the gamma model, may also be interpreted as a tolerance distribution model in which the tolerance distribution is gamma with parameters λ and k. For a discussion of the single-hit model as applied to the high to low dose extrapolation problem, see (9,14); the Report of the Scientific Committee of the Food Safety Council (15,16) and Rai and Van Ryzin (17,18) discuss the application of the multihit model for dose extrapolation.

Other mechanistic models have also been derived from quantitative theories of carcinogenesis. The multistage carcinogenesis theory (19-21) which assumes that a single cell can generate a malignant tumor only after it has undergone a certain number, e.g. k, of heritable changes leads to the multistage model,

$$P(D;\lambda_1,\cdots,\lambda_k) = 1-\exp(-(\lambda_1 D+\lambda_2 D^2+\cdots+\lambda_k D^k)),\quad \lambda_i \geq 0 \quad i=1,\cdots,k$$

The use of this model for extrapolation purposes has been described by Brown (22) and Guess and Crump (23,24).

The multicell carcinogenesis theory of Fisher and Holloman (25) leads to a dose-response function having extrapolation characteristics similar to the multihit model,

$$P(D;\lambda,k) = 1-\exp(-\lambda D^k) \quad , \quad \lambda,k>0$$

This model has also been termed the Weibull model and Van Ryzin (26) discusses its application.

It should be noted that both the single-hit model and the multistage model (when $\lambda_1>0$) become approximately linear at low dose levels. This low dose linearity is an important aspect of "conservative" extrapolation models. Many researchers believe that the true dose-response at low exposure levels is convex, i.e. may have some degree of upward curvature. Therefore, linearity, which represents an "upper bound" to the general class of convex functions, provides conservative extrapolated risk estimates at low doses ("conservative" in the sense of producing higher estimated risks than other convex functions).

Pharmacokinetic Models

Pharmacokinetic hypotheses concerning toxicity from foreign chemicals state that biological effects are manifestations of biochemical interactions between the foreign substances (or substances derived from them) and components of the body. Actual mechanisms of toxicity are many and varied, and the kinetics which relate the concentration and exposure duration of the toxic substance at its site of action with its effect depends upon the mechanism.

A critical problem in the application of pharmacokinetic principles to risk extrapolation is the potential change in metabolism or other biochemical reactions as external exposure levels of the toxic agent decrease. Linear pharmacokinetic models are often used. However, there are numerous examples of nonlinear behavior in the dose range studied, and these nonlinear kinetics pose significant problems for quantitative extrapolation from "high" to "low" doses if the kinetic parameters are not measured (27-29).

Linear kinetics assumes that the reaction rate per unit time, r, of a chemical reaction is proportional to the concentration C of the substance being acted upon, $r \cong kC$; whereas nonlinear kinetics is most often described in the form of a Michaelis-Menten expression, $r \cong aC/(b+C)$, note that for low concentrations, $r \cong (a/b)C$, i.e. linear kinetics, while for high concentrations, $r \cong a$, independent of the concentration C, often referred to as "saturable" kinetics. The parameter a represents the maximum rate of the chemical reaction, and b represents the concentration of the chemical which will produce half this maximum reaction rate.

If all processes are linear, then the concentration rate of the toxic substance at its site of action ('effective dose') will be proportional to the external exposure rate ('administered dose'). However, saturation phenomena may produce different results depending upon the processes affected; if elimination and/or detoxification pathways are saturable, then the effective dose will increase more rapidly with the administered dose than linear kinetics would suggest; if the distribution and/or activation pathways are saturable, then the effective dose will increase less rapidly with the administered dose. These simplified pharmacokinetic models may provide more realistic explanations of observed nonlinear dose-response relationships than other dose-response models currently in use.

Pharmacokinetic models involving nonlinear kinetics of the Michaelis-Menten form have the important extrapolation characteristic of being linear at low dose levels. This low dose linearity contrasts with the low dose nonlinearity of the multihit and Weibull models. Each model, pharmacokinetic, multihit, and Weibull, has the desirable ability to describe either convex (upward curvature) or concave (downward curvature) dose-response relationships. Other models, such as the log normal or multistage, are not consistent with concave relationships. However, the pharmacokinetic model differs from the multihit and Weibull in that it does not assume the nonlinear behavior observed at high dose levels will necessarily correspond to the same nonlinear behavior at low dose levels.

Gehring and Blau (30) and Gehring, et al. (31) discuss this simplified pharmacokinetic model and its extension to more complex reactions with respect to extrapolation of carcinogenic risk from high to low doses. Gehring, et al. (27) applied pharmacokinetic principles to the dose-response of hepatic angiosarcomas in rats exposed to different concentrations of atmospheric vinyl chloride over a period of 12 months. The results of their study are shown in Figure 2. Since the metabolic activation of vinyl chloride appears to be a saturable process, the observed relationship between response, as measured by the proportion of rats with hepatic angiosarcomas, and dose, as measured by the external atmospheric exposure level of vinyl chloride, is clearly nonlinear, showing a leveling out at the highest exposure levels which cannot be explained by a number of the previously discussed dose-response models (e.g. log normal and multistage), but is consistent with a multihit model with $k<1$ 'hits' or a Weibull model with $k<1$ 'stages', both of questionable meaning. However, if dose is measured in terms of the amount of vinyl chloride metabolized, then the dose-response becomes much more linear, and most models provide an adequate fit to the data.

Adjustments for Natural Responsiveness

The mathematical dose-response models described in the preceding sections have assumed responses of the subjects to be due solely to the applied stimuli. However, many toxicity experiments and observational studies show clear evidence that responses can occur even at a zero dose. Thus, any mathematical dose-response function should properly allow for this natural, or 'background', responsiveness.

Two methods have been proposed to incorporate the possibility of response due to factors other than the stimulus in question. The first is commonly known as 'Abbott's correction' which is based on the assumption of an independent action between the stimulus and the background (32). If the probability of response in the absence of any stimulus is denoted by P_0, then the overall response probability at dose level D, assuming independent actions, becomes

$$P(D) = P_0 + (1-P_0)P^*(D)$$

where $P^*(D)$ represents the dose-induced probability of response. The second method assumes that the dose acts in an additive manner with the background environment, producing the overall dose-response model (33)

$$P(D) = P^*(D+D_0)$$

where D_0 represents some unknown background level of the stimulus (or other stimuli that produce the response in a mechanistically dose-additive manner).

It is often difficult to discriminate between the independent and additivity assumption on the basis of dose-response data. Figure 3 compares the theoretical dose-response relationships of these two assumptions where $P^*(D)$ is a log logistic model. The parameters of these models were chosen to minimize their difference. Clearly, a large set of data would be required to determine the proper manner to incorporate background response. To describe the dose-response in this observable response range, this figure shows that this assumption is not an important issue, as both will describe equally data in the observable response range. However, for purposes of low-dose extrapolation, this assumption can have important consequences. Crump, et al. (21) have shown mathematically, that no matter what dose-response model, $P^*(D)$, is used, the additivity assumption will lead to a linear dose-response in the low dose region. This will not necessarily be true for the independent action assumption (note that both assumptions lead to identical mathematical models for overall response rates when the assumed dose-induced model is either the single-hit or multistage). Hoel (34) compares low dose risk extrapolations based on the two assumptions applied to a log

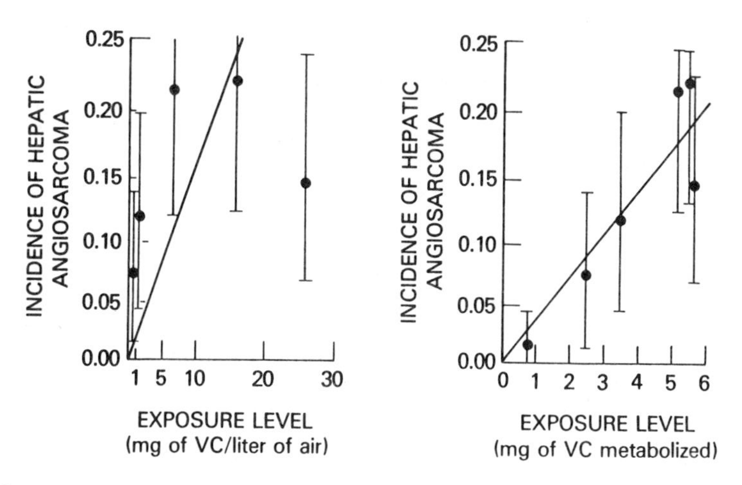

Figure 2. Incidence of Vinyl Chloride induced hepatic angiosarcomas in rats; data from (27).

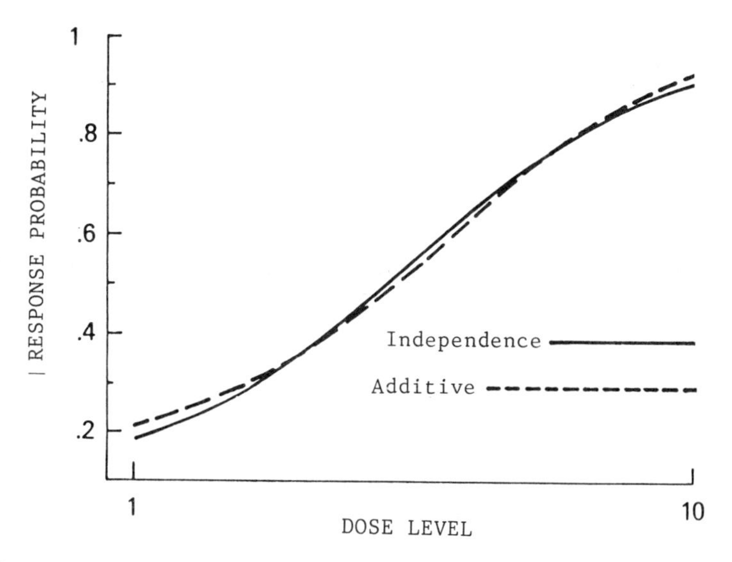

Figure 3. Comparison of log logistic dose-response models assuming independent and additive background.

normal dose-response model. His results are given in Table I. This table clearly shows the low-dose linearity of the additive assumption, and the substantial difference between the additive and independence assumptions at low dose levels. Hoel also examined models which incorporate a mixture of independent and additive background response, and found that low dose linearity prevails except when the background mechanism is totally independent of the dose-induced mechanism.

Table I. Excess Risk P(D)-P(0) for Log Normal Dose Response Model Assuming Independent and Additive Background

Dose (D)	Type of Background: Independent	Additive
10^{0}	4.0×10^{-1}	4.0×10^{-1}
10^{-1}	1.5×10^{-2}	5.2×10^{-2}
10^{-2}	1.6×10^{-5}	5.2×10^{-3}
10^{-3}	3.8×10^{-10}	5.1×10^{-4}
10^{-4}	1.8×10^{-16}	5.1×10^{-5}

*P(0) = 0.1; log normal model slope = 2 from (34)

The existence of a threshold is an important consideration in the evaluation of risk to low levels of environmental toxicants. In this section, the term 'threshold' for a particular toxic response is defined to be that critical level of exposure below which the response attributable to the specific agent is impossible. If there are thresholds, and if they can be quantified, then truely safe levels of a toxic agent can be established.

The existence of thresholds is thought to depend upon the type of toxic effect produced, either a reversible or irreversible effect. Freese (35) discusses the possibility of thresholds for general toxic effects, and more specific teratogenic, mutagenic, and carcinogenic effects. He suggests that many toxic agents inhibit cellular reactions in a reversible manner, and a true threshold may exist if the inhibited reactions do not normally limit the rate of cell metabolism or an organ's function until a certain critical level is attained. However, he believes that thresholds for irreversible mutagenic effects are less likely since the heritable effect upon a single cell may produce untoward effects if the mutated cell replicates.

In discussing thresholds for carcinogenesis, Rall (36) and Brown (37) argue against the existence of a single threshold, but rather that thresholds are likely to vary among members of the population at risk and may be modified by other environmental agents. Mantel, et al. (38) and Brown (37) show mathematically that such population heterogeneity induces an increasing convexity in the population dose-response relationship at low dose levels, and these variable threshold models are difficult to distinguish from nonthreshold convex models. Therefore, when individual thresholds do actually vary within the population, extrapolation of an observed dose-response in order to estimate a population threshold level will, at best, estimate the average threshold of the population at risk. This estimate of the average threshold will have little practical utility since many subjects in the population will have their threshold below this value.

The issue of thresholds for toxic agents is a controversial issue that has yet to be settled. Whether or not a true threshold for a particular toxic effect actually exists may be immaterial as suggested by Mantel (39). A practical threshold can be expected to exist for a variety of reasons. The likelihood of such toxic effects may be affected by dose-dependent rates of absorption, distribution, metabolism, and excretion of the toxic agent. Activation and deactivation may require enzyme reactions that can be induced by the agent itself or some other compound, and cellular repair mechanisms may affect the action of mutagens and carcinogens. However, many researchers suggest use of the no-threshold assumption when extrapolating mutagenic or carcinogenic effects unless knowledge of mechanisms warrant otherwise.

Discrimination among dose-response models

Given a postulated functional form of the dose-response relationship, the frequency of occurrence of toxic effects may be used to estimate the unknown parameters. In addition, this estimated dose-response can be extrapolated to provide either (1) estimates of risk probabilities at lower dose levels, or (2) an estimate of the dose level associated with any particular probability of risk. Implicitly, this approach presumes that the true dose-response can be realized within the postulated functional form used in the estimation and extrapolation procedure. Although this presumption is often not critical for interpolation within the range of observed response rates, it may be extremely critical for extrapolation outside this observable range.

It might be thought that the basis for selection of one particular model over the others would be provided by the observed dose-response. However, this is often not the case, as many dose-response models appear similar to one another over the range of observable response rates. Tables II and III compare the dose-response relationships of the more commonly used models; Table II compares the log normal, log logistic and single-hit models; Table III compares the multihit, Weibull and multistage models.

In the upper panel of Table II, the parameters for these models were chosen to make the response rates equal at dose levels of 1 and 1/4; in the upper panel of Table III, the parameters for the models were chosen to make the response rates equal at dose levels of 2 and 0.5. These tables clearly show that it would take

Table II. Comparison of Dose-Response Relationships Over Range of Observable and Extrapolated Response Rates Log normal, Log logistic, Single-hit models*

Dose Level	Percent Responders		
	Log normal	Log logistic	Single-hit
16	98%	96%	100%
8	93	92	99
4	84	84	94
2	69	70	75
1	50	50	50
1/2	31	30	29
1/4	16	16	16
1/8	7	8	8
1/16	2	4	4
1/100	5×10^{-2}	4×10^{-1}	7×10^{-1}
1/1000	4×10^{-4}	3×10^{-2}	7×10^{-2}
1/1000	1×10^{-7}	2×10^{-3}	7×10^{-3}

* from (51).

an inordinately large set of experimental or observational data to be able to conclude which of the models provide a significantly better fit to an observed dose-response.

If the estimated dose-response is to be used to predict the response rate that would be expected from an exposure level within the range of observable rates, then the models within each of the two sets compared will give similar results. However, extrapolation to exposure levels expected to give very low response rates is highly dependent upon the choice of model, as shown in the lower panels of Tables II and III. These tables extend the dose-response in the upper panels to much lower dose levels. The further one extrapolates from the observable response range, the more divergent the models become. At a dose level which is 1/1000 of the dose giving a 50% response, the single-hit model gives an estimated response rate 200 times that of the lognormal model, and the multistage model gives an estimated response rate over 210 times that of the multihit model.

Krewski and Van Ryzin (40) examined the extrapolation characteristics of six of the more commonly used dose-response models. They applied these models to 20 sets of toxic response data that were taken from the Report of the Scientific Committee of the Food Safety Council (15,16). The toxic responses were both carcinogenic and noncarcinogenic in nature. Of the 19 data sets having an observed convex (i.e. upward curvature) dose-response, all estimates of the virtually safe dose (VSD) at a response rate of $P = 10^{-5}$ or smaller had the ordering, single-hit < multistage < (Weibull, log logistic, multihit) < log normal. That is, the Weibull, log logistic, and multihit produce VSD's of approximately the same order of magnitude, the single-hit model produces the smallest VSD, and the log normal model the largest VSD. In addition, the difference between the extremes, the single-hit and log normal models, is often several orders of magnitude.

Table III. Comparison of Dose-Response Relationships Over Range of Observable and Extrapolated Rates Multihit, Weibull, and Multistage Models

Dose Level	Percent Responders		
	Multihit	Weibull	Multistage
4	99%	99%	100%
3	96	97	98
2	85	85	85
1	50	49	46
0.75	36	35	33
0.50	21	21	21
0.25	7	8	9
0.01	1×10^{-2}	7×10^{-2}	3×10^{-1}
0.001	1×10^{-4}	2×10^{-3}	3×10^{-2}
0.000	1×10^{-6}	7×10^{-5}	3×10^{-3}

Table IV and Figure 4 give an example of this behavior for these models applied to the incidence of liver hepatomas in mice exposed to various levels of DDT (41). This example in Table IV shows that each of the six dose-response models fit the observed data nearly equally well (the multistage model fits the data as well as the others). Therefore, the data in the observable response range (for this study, between 2 and 250 ppm DDT in the daily diet) cannot discriminate among these models. Based on the goodness-of-fit statistics, the Weibull model fits the best ($P = 0.22$), but not significantly better than any of the other models. However, there is a significant difference among the VSD estimated from these models; the log normal model estimates a VSD over 3000 times as large as the single-hit model. Therefore, these experimental data leave the true VSD open to wide speculation. Figure 4 provides a graphical display of these estimated dose-response models over a range of risk levels from 10^{-1} to 10^{-8}. The divergence of these models becomes more apparent as the dose and risk levels become smaller. The analyses of Krewski and Van Ryzin (40) show that this result is a common occurrence.

The fact that an experimental study conducted at exposure levels high enough to give measureable response rates cannot clearly discriminate among these various models, along with the fact that those models show substantial divergence at low exposure levels present one of the major difficulties for the problem of low dose extrapolation. Since the multistage model has the extrapolation characteristics of most other models, Brown (42) has suggested its use to provide estimates of both sampling and model variability for this low dose extrapolation problem.

Table IV. Comparison of Virtually Safe Doses (VSD) Leading to an Excess Risk of 10^{-6} for Various Dose-Response Extrapolation Models (models applied to data from (41))

Extrapolation Model	VSD* (ppm DDT in daily diet)	Goodness-of-fit Statistic of Model to Observed Data χ^2	(d.f.)	P-value
Log normal	6.8×10^{-1}	3.93	(2)	0.14
Weibull	5.0×10^{-2}	3.01	(2)	0.22
Multihit	1.3×10^{-2}	3.31	(2)	0.19
Log logistic	6.6×10^{-3}	3.45	(2)	0.18
Multistage	2.5×10^{-4}		-------**	
Single-hit	2.1×10^{-4}	5.10	(3)	0.16

* 97.5% lower confidence limit on VSD computed by the likelihood method described in (22)

** no goodness-of-fit statistic since the number of parameters equals the number of data points

One would naturally think that since many experimental dose-response studies are conducted with a limited number of animals at each dose level (usually on the order of 100 or fewer) over a range of response rates on the order of 10% - 90%, this problem of wide variation in the VSD might be reduced by testing more animals and using lower dose levels. However, that this will not necessarily be the case is demonstrated by the "megamouse" study of dietary exposure to 2-acetylaminofluorene (2-AAF) conducted at the National Center for Toxicological Research. One of the purposes of this massive study, involving over 24,000 mice, was to describe the carcinogenic dose-response of 2-AAF down to excess risks on the order of 1% (43).

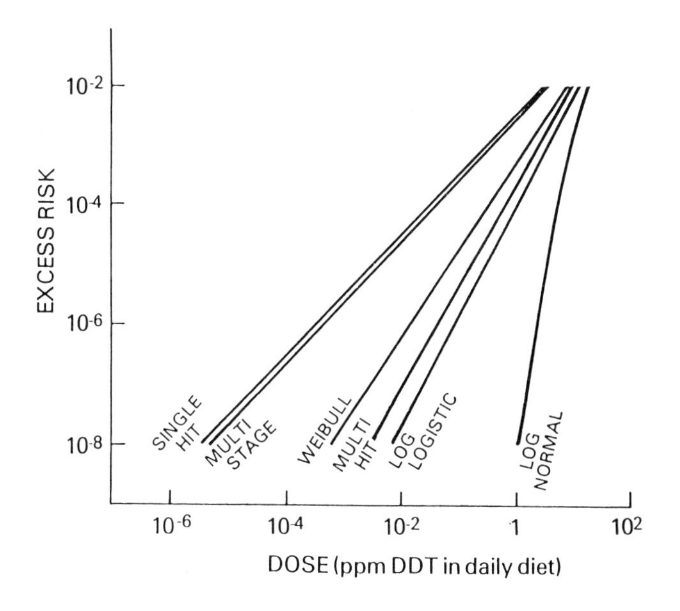

Figure 4. Comparison of high to low dose extrapolation for 6 dose-response models; data from (41).

The results of this study provide an example of the additional information that one might expect to gain by testing large numbers of animals at lower than usual dose levels. The incidence of both bladder and liver neoplasms for those animals which either died naturally or were sacrificed after being exposed for approximately 24 months is shown in Table V (44).

Table V. Incidence of Bladder and Liver Neoplasms In Mice Fed 2-Acetylaminofluorene Continuously 24 Months Following Start of Exposure*

Dose Level (ppm)	Bladder Neoplasms		Liver Neoplasms	
0	2/759**	(0.3%)	20/762	(2.6%)
30	9/2105	(0.4)	164/2109	(7.8)
35	5/1357	(0.4)	128/1361	(9.4)
45	4/881	(0.5)	98/888	(11.0)
60	6/756	(0.8)	118/758	(15.6)
75	13/586	(2.2)	118/587	(20.1)
100	51/297	(17.2)	76/297	(25.6)
150	236/313	(75.4)	126/314	(40.1)

* from (44)
** number mice with neoplasms/number of mice examined

These data were examined to see if the addition of data at dose levels giving low response rates would lead to a reduction in the variation of the VSD estimates. Two extrapolation models, the multistage and the log normal, were applied to these data in a series of calculations. In each case, both models fit the observed data very well. First, the VSD's leading to an excess risk of 10^{-6} are estimated using the controls and the four highest dose groups, 60 - 150 ppm, then the VSD's are reestimated by adding the next lower dose, one at a time. These VSD estimates are shown in Table VI.

These results show that the inclusion of additional low dose data has little effect on the VSD estimates. For bladder neoplasms, which show a highly convex dose-response, the lower confidence limit on the VSD based on the multistage model is increased only 18% (from 3.07×10^{-2} to 3.63×10^{-2}), while that based on the log normal model is hardly changed. For liver neoplasms, which show a nearly linear dose-response, the lower confidence limit on the VSD is increased only 23% for the multistage model and is decreased for the log normal model. The differences in the VSD estimates from these two extrapolation models is little affected by these additional low dose data: for bladder neoplasms, the additional data decreases the difference from a log normal/multistage ratio of 1120 ($34.4/3.07 \times 10^{-2}$) to 950 ($34.5/3.63 \times 10^{-2}$); for liver neoplasms this ratio is reduced from 1380 to 890. Therefore, these additonal low dose data, based on substantial numbers of animals, has little effect on the VSD estimates for a particular extrapolation model, and, more importantly, has little effect on reducing the variation in VSD estimates between different models.

Table VI. Virtually Safe Doses (VSD) for 2-AAF Based Multistage and Log Normal Models Applied To Different Dose Level Combinations

Dose Levels Used (ppm)	VSD (ppm)* Bladder Neoplasms Multistage	Bladder Neoplasms Lognormal	Liver Neoplasms Multistage	Liver Neoplasms Lognormal
0, 60 - 150	3.07×10^{-2}	34.4	3.50×10^{-4}	4.84×10^{-1}
0, 45 - 150	4.12×10^{-2}	34.5	3.82×10^{-4}	5.27×10^{-1}
0, 35 - 150	4.48×10^{-2}	34.5	3.99×10^{-4}	4.03×10^{-1}
0, 30 - 150	3.63×10^{-2}	34.5	4.32×10^{-4}	3.84×10^{-1}

* 97.5% lower confidence limit on VSD leading to an excess risk of 10^{-6} computed by the likelihood method described in (22)

For situations of long-term chronic exposure to a toxic agent, the relationship of risk to the rate and duration of exposure is often of importance when estimating risk for different exposure situations. One commonly employed assumption is that risk is dependent upon total cumulative exposure. Thus, for example, an individual exposed daily to 1 mg of the toxic agent for 20 years duration is assumed to have the same risk as another individual exposed daily to 10 mg for 2 years duration. Again the mechanism of toxic action will determine the validity of this assumption. For example, if the toxic agent accumulates at the target site, and the response becomes evident when the accumulated level attains some critical level, then this total dose assumption may be warrented. However, physiological processes, such as detoxification or elimination from the target site, are likely to be dependent upon the accumulated level, and thus may modify this simple total dose relationship.

The multistage theory of carcinogenesis predicts that cancer risk is dependent upon the dose rate and duration of exposure, but not necessarily leading to a relationship with total dose, the product of rate and duration. Whittemore and Keller (45), Whittemore (46), and Day and Brown (47) discuss these multistage theories and indicate that the risk of cancer is likely to be the product of two different functions of dose rate and duration. In an analysis of cigarette smoking and lung cancer, Doll (48) found lung cancer incidence rates rise as approximately the fourth power of duration while Doll and Peto (49) found that incidence rises as approximately the first or second power of daily number of cigarettes smoked. Besides being a function of both dose rate and duration, the multistage theory also predicts that cancer risk may be a function of the age at which exposure first begins and the amount of time following cessation of exposure. Whittemore (46) and Day and Brown (47) show how risk may be a function of these two factors, depending upon the stage of the carcinogenic process affected by the toxic agent. For example, exposure at a young age to a carcinogen affecting an early stage (i.e. an initiator) is predicted by the theory to have a greater effect on future cancer risk than the same exposure at a later age. The converse is predicted for exposure to a carcinogen which affects a late stage in the process.

Therefore, the multistage theory of carcinogenesis predicts another level of complexity in extrapolating cancer risk from one exposure situation to another, since a limited duration (e.g. 10 years) of exposure to the same dose rate will not necessarily produce the same excess cancer risk in two otherwise identical individuals whose exposure period is during different ages of their life.

Summary and Conclusions

The preceeding sections have discussed the general problem of high dose to low dose extrapolation within a single animal species. The purpose of this extrapolation is to estimate the effects of low level exposure to toxic agents known to be associated with undesired effects at high dose levels.

Mathematical models of dose-response are necessary for this extrapolation process since the low dose effects, expected to be on the order of response rates of 10^{-6}, are too small to be accurately measured with limited study sample sizes. A number of mathematical dose-response models have been proposed for extrapolation purposes; we previously saw how similar they can appear to one another in the range of observable response rates, yet how different they become at lower, unobservable response rates, the region of primary interest. This is the single, most important limitation of this extrapolation methodology. An estimate of risk at a particular low dose, or an estimate of the dose leading to a prespecific level of risk is highly dependent upon the mathematical form of the presumed dose-response; Section IV shows that differences of 3 - 4 orders of magnitude are not uncommon. The proposal of "new" models, unless based upon strong mechanistic information, will not alleviate the difficulties. A Bayesian approach, along the lines suggested by Altshuler (50), might be a method by which prior judgements about the plausibilities of different functional forms, in the light of toxicological and biological information, could be incorporated into the extrapolation process. The contribution from statisticians and model-builders has reached an impass, and more accurate extrapolations are not possible without additional information on the mechanisms of action of the toxic agents.

Pharmacokinetic information on the fate of a toxic agent once it enters the body is beginning to be incorporated into the high to low dose extrapolation process. Nonlinear kinetics may be an important determinant of the nonlinear dose-response relationships often observed in experimental studies of toxic agents. As noted in previously, Gehring et al (27) have shown that the metabolism of inhaled vinyl chloride is a saturable process that provides one explanation of the concave liver carcinogenesis dose-response observed in animal studies. In a study of urethane-induced pulmonary adenomas, White (28) showed that the convex relationship between the amount of urethane injected into a mouse lung and the number of subsequent lung adenomas could be explained by nonlinear kinetics of excretion. Such pharmacokinetic models and dose-response studies of the kinetics of physiological processes might considerably strengthen the ability to extrapolate from high to low dose levels. This avenue of investigation holds potentially great promise for the future.

Other sources of uncertainty in high to low dose extrapolation include: (1) the possible existence of thresholds; (2) heterogeneity of sensitivity to the toxic agent among members of the exposed population; and (3) mechanisms of action for carcinogens (i.e. whether the agent initiates the process or acts at a later stage). The existence of a single threshold for the entire exposed population should allow for estimation of a clearly safe level of exposure. However, its estimation could be associated with a high degree of uncertainty. Heterogeneity in individual thresholds and sensitivity to the toxic agent induces additional uncertainty in high to low dose extrapolations. The relationship of dose rate and duration of exposure discussed in Section IV indicates that similar exposure patterns (i.e. same dose rate and duration) do not necessarily lead to similar levels of risk. Thus, uncertainty in the mechanism of toxic action induces another potentially large uncertainty into risk extrapolations.

Therefore, all these sources of uncertainty, (1) dose-response model, (2) pharmacokinetic behavior of the toxic agent, (3) thresholds, (4) heterogeneity, and (5) mechanisms of action, lead to potentially enormous variation in estimates of risk from high to low dose extrapolations.

Literature Cited

1. Schneiderman, M.A.; Mantel, N.; Brown, C.C. Ann. NY Acad. Sci. 1975, 246, 237-246.
2. Gaddum, J.H. Medical Research Council, Special Report Series No. 183, 1933.

3. Bliss, C.I. Science 1934, 79, 38-39.
4. Bliss, C.I. Ann. Appl. Biol. 1935, 22, 134-167.
5. Finney, D.J. "Probit Analysis, 3rd Ed." Cambridge University Press: London, 1971.
6. Mantel, N.; Bryan, W. J. Natl. Cancer Inst. 1961, 27, 455-470
7. Mantel, N.; Bohidar, N.; Brown, C.; Ciminera, J.; Tukey J. Cancer Res. 1976, 35, 865-872.
8. Schneiderman, M.A. J. Wash. Acad. Sci. 1974, 64, 68-78.
9. Hoel, D.G.; Gaylor, D.; Kirschstein, R.; Saffiotti, U.; Schneiderman, M. J. Tox. and Environ. Health 1975, 1, 133-151.
10. Hartley, H.O.; Sielken, R.L. Biometrics 1977, 33, 1-30.
11. Worcester, J.; Wilson, E.B. Proc. Natl. Acad. Sci. 1943, 29, 78-85.
12. Berkson, J. J. Amer. Stat. Assn. 1944, 39, 134-167.
13. Turner, M. Math. Biosci. 1975, 23, 219-235.
14. "The Effects of Populations of Exposure to Low Levels of Ionizing Radiation," National Academy of Sciences, 1980.
15. Scientific Committee, Food Safety Council. Food and Cosmetic Tox. 1978, 16 supplement 2, 1-136.
16. Scientific Committee, Food Safety Council "Proposed System for Food Safety Assessment"; Food Safety Council: Washington, D.C., 1980; p. 137.
17. Rai, K.; Van Ryzin, J. in "Energy and Health"; Breslow, N.; Whittemore, A. Eds.; SIAM: Philadelphia, 1979; p. 99.
18. Rai, K. and Van Ryzin, J. Biometics 1981, 37, 341-352.
19. Armitage, P; Doll, R. Brit J. Ca. 1954, 8 1-12.
20. Armitage, P; Doll, R. in "Proceedings of the Fourth Berkeley Symposium on Mathematical Statistics and Probability (Vol. 4)"; Neyman, J. Ed.; University of California Press: Berkely and Los Angeles, 1961, p. 19.
21. Crump, K.S.; Hoel, D.; Langley, C.; Peto, R. Cancer Res. 1976, 36, 2973-2979.
22. Brown, C. J. Natl. Cancer Instit. 1978, 60, 101-108.
23. Guess, H.A.; Crump, K.S. Math. Biosci. 1976, 32, 15-36.
24. Guess, H.A.; Crump, K.S. Environ. Health Perspect. 1978, 22, 149-152.
25. Fisher, J.C.; Holloman, J.H. Cancer 1951, 4, 916-918.
26. Van Ryzin, J. J. Occup. Med. 1980, 22, 321-326.
27. Gehring, P.J.; Watanabe, P.G.; Park, C.N. Tox. App. Pharm. 1978, 44, 581-591.
28. White, M. in "Proceedings of the Sixth Berkely Symposium on Mathematical Statistics and Probability, Vol. 4"; Lecam, L.E.; Neyman, J.; Scott, E.L. Eds;, University of California Press: Berkeley and Los Angeles, 1972; p. 287.
29. Hoel, D.G.; Kaplan, N.L.; Anderson, M.W. Science 1983, 219, 1032-1037.
30. Gehring, P,J.; Blau, G.E. J. Envir. Path and Tox. 1977, 1, 163-179.

31. Gehring, P.J.; Watanabe, P.G.; Young, J.D. in "Origins of Human Cancer (Book A: Incidence of Cancer in Humans)"; Hiah, H.H; Watson, J.D.; Winston, J.A. Eds.; Spring Harbour Laboratory: Cold Springs Harbour, 1977; p. 187.
32. Abbott, W.S. J. Econ. Ent. 1925, 18, 265-267.
33. Albert, R; Altshuler, B. in "Radionuclide Carcinogenesis"; Ballou, J.; Mahlum, D.; Sanders, C. Eds.; AEC Symposium Series, Conf-720505, 1973, p. 233.
34. Hoel, D.G. Fed. Proceed. 1980, 39, 67-69.
35. Freese, E. Environ. Health Perspectives, Experimental Issue 6, 1973, 171-178.
36. Rall, D.P. Environ. Health Perspect. 1978, 22, 163-165.
37. Brown, C. Oncology 1976, 33, 62-65.
38. Mantel, N., Heston, W.; Gurian, J. J. Natl. Cancer Instit. 1961, 27, 203-215.
39. Mantel, N. Clin. Pharm. and Ther. 1963, 4, 104-109.
40. Krewski, D.; Van Ryzin, J. in "Current Topics in Probability and Statistics"; Sorgo, M.; Dowson, D.; Rao, J.N.K.; Saleh, E. Eds.; North- Holland: New York, 1981; p. 201.
41. Tomatis, L.; Turusov, V.; Day, N.; Charles, R.T. Int. J. Cancer 1972, 10, 489-506.
42. Brown, C.C. Environ. Health Perspect. 1978, 22, 183-184.
43. Cairns, T. J. Environ. Path. and Toxicol. 1980, 3, 1-8.
44. Farmer, J.H.; Kodell; R.L.; Greenman, D.L.; Shaw, G.W. J. Environ. Path. and Toxicol. 1980, 3, 55-68.
45. Whittemore, A.S.; Keller, J.B. SIAM Review 1978, 20, 1-30.
46. Whittemore, A.S. Am. J. Epidem. 1977, 106, 418-432.
47. Day, N.E.; Brown, C.C. J. Natl. Cancer Instit. 1980, 64, 977-989.
48. Doll, R. J. Royal Stat. Soc. A 1971, 134, 133-155.
49. Doll, R.; Peto, R. J. Epidem. Community Health, 1978, 32, 303-313.
50. Altshuler, B. in "Environmental Health, Quantitative Methods"; Whittemore, A. Ed.; Society for Industrial and Applied Mathematics, Philadelphia, 1977; p. 31.
51. Food and Drug Advisory Committee on Protocols for Safety Evaluation Tox. App. Pharm. 1971, 20, 419-438.

RECEIVED October 20, 1983

MANAGEMENT OF CHEMICAL RISKS

6

Legal Considerations in Risk Assessment Under Federal Regulatory Statutes

PETER BARTON HUTT

Covington & Burling, 1201 Pennsylvania Avenue, N.W., Washington, DC 20044

Throughout history, regulatory statutes to protect the public health and safety have been worded in sufficiently broad and general terms to authorize the government to utilize current scientific knowledge in determining adequate public protection. The statutory requirements of current health and safety laws implemented by FDA, EPA, CPSC and OSHA are sufficiently flexible to allow the adoption of whatever analytical and decision-making methodology best represents the public interest. Implementation of current regulatory statutes in this field is therefore constrained largely by the current state of scientific knowledge rather than by rigid or obsolete statutory requirements.

Government regulation to protect the public health and safety is not a recent phenomenon. It has been prevalent throughout recorded history. Numerous Federal statutes now exist to protect the safety of the products we consume and use, and all aspects of our environment.

Neither legislatures nor the courts have ever enunciated an operational definition of "safety." Implementation of Federal regulatory requirements relating to safety has therefore been delegated to those who must comply and to those who must enforce compliance. Over the years, gradually more sophisticated measures of safety have evolved through advances in the biological sciences. Fortunately, long-established principles of statutory interpretation and

0097–6156/84/0239–0083$06.00/0

administrative discretion permit wide latitude to adopt these new approaches in the implementation of safety requirements. Thus, implementation of Federal regulatory statutes in the health and safety field is constrained by the current state of scientific knowledge rather than by rigid or obsolete statutory requirements.

The history of health and safety regulation reflects a continuing evolution of assessment of health risks from the chemicals in our environment -- whether those chemicals are in food, drugs, other consumer products, the workplace, or the air that we breath and the water that we drink. As science has become more refined, so has safety/risk assessment. Regulators throughout history have used the best form of safety/risk assessment available at any particular time, but no one has ever believed, or believes now, that an adequate safety/risk assessment methodology has been found. In this field, as in most other fields of scientific endeavor, there is still a great deal to be learned.

This paper traces some of the relevant statutory and regulatory history in the field of health and safety. It discusses legal principles involved in use of safety/risk assessment under current Federal statutes.

Background History

Government regulation of health and safety is perhaps the oldest form of government regulation of commercial activity. Regulation of food, in particular, has been found in all recorded civilizations. Over many centuries, health and safety regulation has expanded to other consumer products and to the various elements of our environment.

English precedent is particularly useful, both because it is relatively well-recorded and because it has formed the basis for much of our own early regulatory effort.

King John's Assize of Bread, first promulgated in 1203, initially began as purely economic regulation, but soon expanded to protect staple foods against any form of adulteration. The first recorded version of this law, in 1266, prohibited "contagious flesh" and any wine, flesh, or fish "not wholesome for Man's Body."(1) During the next 600 years, the English Parliament enacted numerous laws prohibiting the adulteration of specific foods.

Frederick Accum's classic treatise on adulterations of food and drugs, published in 1820, was the first popular book to focus public attention on the seriousness of the problem.(2) Accum pointed out that some adulterations were harmless, resulting merely in fraud on the consumer, but others were deleterious to health as well. After widely-publicized investigations of food and drug adulteration, England enacted national statutes in 1860 and 1875 prohibiting any ingredient "injurious to health."(3-4)

Concurrent with enactment of these statutory provisions, the English judiciary was evolving a parallel body of both civil and criminal common law. Causes of action were recognized against those who sold adulterated food and those who maintained any common nuisance.(5)

The English emigrants who populated the American Colonies thus brought to this country a coherent and consistent tradition of government protection of the public health. The Colonies followed this precedent closely, both before and after they achieved independence.

The first general food adulteration statute in the world, indeed, was enacted by Masschusets in 1785.(6) It prohibited "diseased, corrupted, contagious, or unwholesome provisions" of any kind. Following Shattuck's landmark 1850 report on the importance of sanitation to public health(7), there was a veritable explosion of state public health laws, including laws regulating food and drugs, in the last half of the 19th century.(8)

Federal regulatory laws, enacted by Congress to preserve the public health, were enacted much earlier in the history of our country than most people realize. The 1813 Vaccine Act authorized the federal government to determine what vaccines were "genuine" in order to disseminate them to the public.(9) In 1848, Congress prohibited the importation of any drug that was "improper, unsafe, or dangerious."(10) The 1886 Oleomargarine Act prohibited any ingredients "deleterious to public health."(11)

Our modern era of food and drug regulatory law began with enactment of the Biologics Act of 1902 (12), the Federal Food and Drugs Act of 1906 (13), and the Federal Meat Inspection Act of 1906.(14-15) These laws were intended to insure that biologics would "yield their intended results" and that food would contain no "added poisonous or other added deleterious ingredient which may render such article injurious to health."

Other regulatory laws enacted by Congress in the first decade of this century were also concerned with public health and safety. The Transportation of Explosives Act of 1909 required the ICC to promulgate regulations "for safe transportation of explosives."(16) The Insecticides Act of 1910 prohibited any adulteration of pesticide products.(17)

Since then, our national laws have been expanded to cover additional subjects, and obviously have been made far more detailed and complex. It is questionable, however, whether the substance of those early statutes has been significantly changed. Both the statutory language used by Congress in those early years, and the judicial precedent interpreting it, suggest that there was as much authority then as there is now to assure appropriate government health and safety regulation.

Regulatory Control Mechanisms In Federal Health And Safety Statutes

Both the early Federal regulatory statutes and the more recent ones employ a wide variety of control mechanisms. These include private enforcement, government policing of the marketplace, development of voluntary and mandatory standards, and various forms of premarket notification, testing, and approval. These different mechanisms reflect the degree to which public and Congress wish to assure safety. As public concern about safety/risk rises, gradually more stringent statutory controls are imposed. Put another way, these mechanisms determine the cost of reducing public uncertainty and concern about safety/risk.

None of these mechanisms, however, determines to any significant degree how a regulator decides what is and is not safe. That decision is governed by the safety/risk standard adopted by Congress in each individual statute.

Safety/Risk Standards In Federal Health And Safety Statutes

It is very difficult to classify the various safety/risk standards in Federal health and safety statutes, for a number of interrelated reasons. Congress seldom uses exactly the same statutory language twice. It is often not feasible to determine whether these differences in language represent simply historical accident or a true congressional intent to convey a different meaning. Nor is the intended meaning of the statutory language ever clear from the statute itself. Quite frequently, Congress resolves controversy over such matters by very general language or simply by avoiding the issue. Finally, for many of these statutes there is relatively little judicial interpretation. Even when courts must wrestle with statutory language, the unique facts of the case often result in a decision that provides no helpful precedent.

In general, however, there are three broad classes of safety/risk standards that have been adopted in Federal statutes: (1) those that relate solely to safety/risk, (2) those that relate both to safety/risk and to other societal consequences of regulation as well, and (3) those that deal with particular issues, such as a specific substance or type of toxicity.

Safety/risk standards. There are many examples throughout history of this type of statutory safety/risk standard. As early as 1266, England adopted a statutory standard of "not wholesome for man's body."(1) In its 1875 law, some 600 years later, it adopted a standard of "injurious to health."(4) These two standards are, of course, indistinguishable from both a legal and a practical standpoint.

Under American law, the experience has been quite similar. The Massachusets statute of 1785 adopted a standard of "diseased, corrupted, contagious, or unwholesome" substances.(6) The 1886 margarine statute used a standard of "deleterious to public health."(11) Both in 1906 and in 1938, Congress prohibited any added "poisonous or deleterious substance" that may render the food "injurious to health."(13,18)

The term "safe" has been adopted as a statutory standard only relatively infrequently. The 1909 Transportation of Explosives Act required that explosives be transported in a "safe" manner (16), and the Federal Food, Drug, and Cosmetic Act (FD&C Act) of 1938 required that all "new drugs" be shown to be "safe."(19) Legislation enacted during 1954-1968 to regulate pesticide residues on food, food additives, color additives, and animal drugs, also required proof of "safety."(20-23) As will be discussed in greater detail below, however, this term has been virtually abandoned in all Federal health and safety legislation enacted since 1970.

In only two instances has a safety/risk standard been interpreted and applied as an absolute, rather than a relative, standard. Beginning in 1950, FDA interpreted the "may render injurious to health" standard as an absolute prohibition of any carcinogenic substance in food.(24-25) This interpretation was based upon the conclusion that current scientific knowledge did not permit the determination of any safe level of a carcinogen. FDA also concluded in the 1950s that the statutory requirement that coal tar colors for use in food must be "harmless" meant that they must be shown to be harmless per se, and not merely harmless in the quantity used in the food. This interpretation was upheld by the Supreme Court in 1958 (26) and ultimately led to enactment of the Color Additives Amendment of 1960, which required that color additives only be shown to be "safe" under their conditions of use.(22)

In all instances other than these two exceptions, safety/risk standards have uniformly been interpreted to incorporate a standard of relative safety/risk under actual conditions of use. Nowhere is this clearer than in the statutory requirement that all new drugs be proved "safe," a requirement that has existed from 1938 to this day.

The definitive interpretation of these various safety/risk standards was established by the Supreme Court in the 1914 Lexington Mill decision.(27) FDA contended that the "injurious to health" standard required the Agency only to show that a food ingredient was not harmless per se. Industry contended, in contrast, that FDA must show actual deleterious effects in humans before the statutory standard was met. The Supreme Court agreed with neither party.

The Court stated that FDA need show only a reasonable possibility of a deleterious effect in order to satisfy this statutory standard. The Agency need not prove actual harm to humans. On the other hand, the Court concluded that this reasonable possibility must be determined in relation to the quantity of the substance actually used in food. Thus, the standard was not absolute harmlessness per se, but rather the reasonable likelihood of harmlessness under conditions of use.

As thus articulated by the Supreme Court almost 70 years ago, a food ingredient is unlawful if FDA can show a reasonable possibility of harm under actual conditions of use. That standard has since been explicitly adopted by Congress in the legislative history of the FD&C Act in 1938 (28-29), reaffirmed by the Supreme Court (26), and applied consistently by lower courts throughout the country for many years.(30)

Indeed, it has remained the major bulwark for food safety protection even in the two decades following enactment of the special laws governing food additives and color additives. Both FDA and industry have uniquely remained united behind this standard, resisting any effort to revise it.

Not surprisingly, the Supreme Court adopted the same approach in interpreting the Occupational Safety and Health Act in the landmark Benzene case in 1980.(31) The Court concluded that the OSH Act prohibits only a significant risk to health, and does not impose a standard of absolute safety.

Standards balancing safety/risk against other societal interests. The second general type of safety/risk standard contained in Federal health and safety statutes requires consideration both of safety/risk and of other societal consequences of regulation. In practice, of course, all regulatory statutes are administered this way. FDA has quite understandably declined to ban essential nutrients shown to be carcinogenic in tests where, if the same results had been obtained on trivial ingredients, there would have been an instantanious ban.(32)

Statutory provisions requiring a balancing of safety/risk against other considerations have existed for many years. Section 406 of the FD&C Act, for example, has provided since 1938 that FDA may establish tolerances for poisonous or deleterious substances, that would otherwise be banned, if those substances are required in the production of food or are unavoidable under good manufacturing practices. It is this provision under which FDA has established a permissible level of aflatoxin in peanuts, corn, milk, and other agricultural commodities.(33)

Since 1970, virtually all federal health and safety statutes have adopted a general safety standard of "unreasonable risk," and have required consideration by regulatory agencies of the broad societal consequences of regulatory action. Such statutes as the Consumer Product Safety Act of 1972 (34), the Federal Environmental Pesticide Control Act of 1972 (35), and the Toxic Substances Control Act of 1976 (36), adopt this approach. These statutes make no attempt to define the critical phrase "unreasonable risk," just as the earlier statutes made no attempt to define such terms as "injurious" or "safe." Some do contain a laundry list of factors to be considered in determining regulatory action, such as the effect on the economy and small business, but they provide little substantive guidance for specific regulatory decisions.

This approach calls upon the Agency to take into account the level of risk and extent of possible reduction of that risk, the beneficial impact this will have upon the public, the various societal costs of the regulation, the detrimental impact that these will have, and other related factors. In practice, however, this kind of rigorous analysis rarely, if ever, occurs.

The uniform approach of agencies and courts faced with this standard is to avoid the benefit or cost issues and to focus upon the question of safety/risk. This derives from the practical fact that no one has yet determined how to balance benefits or costs against risks. Thus, in the *Slide 'N' Dive* decision, the court remanded a consumer product safety standard containing a warning label to the Consumer Product Safety Commission for reconsideration because the risk addressed by the warning was no greater than the risk of any citizen being hit by lightning.(37) In the numerous pesticide cases, EPA has focused almost exclusively on the level of risk.(38) If the level of risk was insignificant, the agency did not act. If the level of risk was substantial, the pesticide was banned. The benefits and costs of regulatory action were discussed at length, to satisfy the statutory requirement, but EPA has never enunciated an operational methodology for balancing those elements or even purported to make a decision based upon such a delicate analysis.

In practice, therefore, the distinction between these first two types of statutory safety/risk standards has not only been blurred, but probably obliterated. Both focus on the safety/risk element of the issue. Both also consider the benefit/cost element, largely *sub silentio*, in a broad brush and common sense way. If regulatory action is simply absurd, and will make the agency look foolish, it will be abandoned. But close analysis of costs and benefits has had no significant impact upon any of these regulatory decisions.

Standards reflecting specific congressional safety determinations. The third general type of safety/risk standard included in Federal health and safety statutes involves highly particularized congressional determinations designed to direct specific regulatory decisions. These have typically been of two different types, relating to (1) particular forms of toxicity and (2) specific named substances.

The best example of a specific statutory safety/risk standard for a particular form of toxicity is the Delaney Clause.(39) This example also illustrates why this type of statutory enactment is relatively ineffective in directing agency action.

As virtually everyone knows, the Delaney Clause prohibits the addition to food of any additive that has been found to induce cancer upon ingestion by test animals. As very few people realize, however, that general proposition is riddled with statutory exemptions, has been subjected to constant administrative exceptions created by FDA to avoid absurd results, and thus has been invoked only twice in its 24-year existence.(25) It simply is not an important factor in FDA's decision-making on food safety. If it were repealed tomorrow, and the other food safety statutory provisions remained unchanged, not a single FDA decision during the past 24 years would be changed.

There have also been a number of specific safety/risk standards or exceptions adopted by Congress for particular substances. For example, the Toxic Substances Control Act of 1976 explicitly required EPA to take regulatory action to remove PCBs from the environment.(40) Even in the face of the seemingly clear and direct congressional mandate, however EPA has found ample room for reasonable interpretation. The exceptions created by EPA have provoked substantial administrative confrontation and court litigation, but have in principle been upheld in the courts.(41)

Thus, even with the most rigid statutory provision, directed either to specific forms of toxicity or individual substances, regulatory action has inevitably been determined on the basis of the same basic safety/risk standard embodied in the first two types of general standards already discussed, tempered by sound common sense.

Principles Of Statutory Construction

Federal health and safety regulatory statutes have been interpreted and applied in highly flexible and common sense ways largely because of the existence of a number of important rules of statutory construction. These rules have been created by the judiciary over many decades, and in some instances centuries, as part of our unwritten common law. They do not depend on the words of a particular statute, or the intent of

Congress as expressed in legislative history. They exist for all time and for all statutes.

Three examples will illustrate the importance of these principles. Courts everywhere have recognized the maxim that de minimus non curat lex, the law does not concern itself with trifles. Courts have relied upon this principle in interpreting the Food Additive Amendments (42), the Clean Air Act (43), and the PCB provisions in the Toxic Substances Control Act.(41) The Supreme Court's decision that the OSH Act only bans a significant risk undoubtedly rests upon this fundamental principle as well.

A related principle is that courts will interpret a statute to achieve its broad purpose, and to preclude absurd or futile results.(44-46) Similarly, courts have stated that regulatory agencies have inherent nonstatutory authority to rationalize a statute by creating administrative exemptions even where the statute does not otherwise provide for them.(47)

Each of these principles thus expands a regulatory agency's discretionary authority to adopt common sense solutions to the difficult safety/risk issues that it faces.

Comparison Of Federal Regulatory Statutes

In light of this analysis, it is apparent that the differences among Federal health and safety statutes are more apparent than real. All of these statutes rely essentially upon a very broad and generalized safety/risk standard. The statutory safety/risk standards contained in existing Federal regulatory statutes neither require nor preclude any particular approach to regulation. Agencies clearly possess the administrative discretion to expand or contract their legislative mandates in numerous ways. A seemingly rigid statute like the Delaney Clause can readily be made flexible, and a seemingly flexible standard can readily become rigid, simply through administrative interpretation and implementation.

The real difference among these statutes thus exists in the agencies themselves, not in the specific statutory terms. The history and tradition of an agency, the nature of its constituency, the characteristics of its personnel, and other similar factors truly determine its approach to regulation far more directly than any statutory language.

Use Of Safety/Risk Assessment In Regulatory Decisions

All health and safety regulation depends upon some form of safety/risk assessment. That assessment may be either qualitative or quantitative in nature. But no regulatory action can be taken in the field of health and safety without some assessment of the safety/risk involved.

For centuries, regulatory activity relied solely upon expert judgment regarding safety/risk. During the past three decades, it has increasingly relied upon quantitative safety/risk estimates, involving such elements as safety factors, the limits of detection methodology, and extrapolation from animal feeding studies to levels of insignificant risk.

It is a truism that, at any particular time in history, the approaches to safety/risk assessment then in use are not perfect. It is equally true that, with time, they will improve. Regulatory decisions, on the other hand, cannot await development of a perfect, or even a better, methodology for assessing safety/risk. They must be made at that time, with the best means currently available.(48)

Thus, a regulator must use the best available approach at that moment to analyze toxicity data and to make a reasonable regulatory decision. Statutory language is not dispositive of this matter, and is not even helpful. It offers no direction to the regulator. Regardless of the specific statutory safety/risk standard invoked, the safety/risk assessment approach to be used in any particular situation, at any particular time, is necessarily left to the judgment and discretion of the agency.

At this particular moment, it appears incontestable that quantitative safety/risk assessment methodology based upon extrapolation from animal feeding studies presents a decided improvement in guiding regulatory decisions over earlier safety/risk assessment approaches, such as expert judgment or the use of arbitrary safety factors. This does not mean that such extrapolations are accurate or should be the determining factor in all regulatory decisions. It does mean that this approach represents a better way to analyze toxicity data in order to make it relevant to the regulatory decisions that must be made. That is why FDA has begun to use it so extensively in its daily regulatory work.(49-60) Ten years from now we undoubtedly will look back upon the current extrapolation methods as obsolete, but today they represent the best that we have.

There is no statutory prohibition, in any of the current Federal health and safety laws, in making daily regulatory decisions utilizing quantitative safety/risk assessment based upon extrapolation from animal toxicity studies. Indeed, the court cases already discussed, from 1914 to the present, lead inexorably to this approach.

Literature Cited

1. 51 Hen. III, st. 6 (1266).
2. Accum, F. "A Treatise on Adulterations of Food and Culinary Poisons"; 1820.
3. 23 & 24 Vict., c. 84 (1860).
4. 38 & 39 Vict., c. 63 (1875).
5. Hutt, P.B., Food Drug Cosm. L.J. 1960, 31, 246.
6. Mass. Act of March 8, 1785, Food Drug Cosm. L.J. 1976 31, 246.
7. Shattuck, L. "Report of the Sanitary Commission of Massachusetts"; 1850.
8. Wiley, H. "Officials Charged with the Enforcement of Food Laws in the United States and Canada", USDA Circ. 16, 1904.
9. 2 Stat. 806 (1813).
10. 9 Stat. 237 (1848).
11. 24 Stat. 209 (1886).
12. 32 Stat. 728 (1902).
13. 34 Stat. 768 (1906).
14. 34 Stat. 669, 674 (1906).
15. 34 Stat. 1256, 1260 (1907).
16. 35 Stat. 554 (1909).
17. 36 Stat. 331 (1910).
18. 52 Stat. 1040 (1938), 21 U.S.C. 342(a)(1).
19. 52 Stat. 1040 (1938), 21 U.S.C. 355.
20. 68 Stat. 511 (1954), 21 U.S.C. 346a.
21. 72 Stat. 1784 (1958), 21 U.S.C. 348.
22. 74 Stat. 397 (1960), 21 U.S.C. 376.
23. 82 Stat. 342 (1968), 21 U.S.C. 360b.
24. 15 Fed. Reg. 321 (January 19, 1950).
25. Hutt, P.B. Food Drug Cosm. L.J. 1978, 33, 541, 543.
26. Flemming v. Florida Citrus Exchange, 358 U.S. 153 (1958).
27. United States v. Lexington Mill & Elevator Co., 232 U.S. 399 (1914).
28. S. Rep. No. 493, 73d Cong., 2d Sess. 3 (1934).
29. S. Rep. No. 361, 74th Cong., 1st Sess. 6 (1935).
30. United States v. 2,116 Boxes of Boned Beef, 516 F.Supp. 321 (D.Kan. 1981).

31. Industrial Union Department, AFL-CIO v. American Petroleum Institute, 448 U.S. 607 (1980).
32. Citizen Petition submitted by the Grocery Manufacturers of America to the Food and Drug Administration 32-41 (August 1981).
33. 39 Fed.Reg. 42748 (December 6, 1974).
34. 86 Stat. 1207 (1972), 15 U.S.C. 2051.
35. 86 Stat. 973 (1972), 7 U.S.C. 136.
36. 90 Stat. 2003 (1976), 15 U.S.C. 2601.
37. Aqua Slide 'N' Dive Corp. v. CPSC, 569 F.2d 831 (5th Cir. 1978).
38. Cooper, R. Food Drug Cosm. L.J. 1978, 33, 755, 762-763.
39. 21 U.S.C. 348(c)(3)(A).
40. 15 U.S.C. 2605(e).
41. Environmental Defense Fund v. EPA, 636 F.2d 1267 (D.C. Cir. 1980).
42. Monsanto Co. v. Kennedy, 613 F.2d 947, 954-955 (D.C. Cir. 1979).
43. Alabama Power Co. v. Costle, 636 F.2d 323, 357-360 (D.C. Cir. 1979).
44. E.g., Sierra Club v. Train, 557 F.2d 485, 490 (5th Cir. 1977).
45. Quinn v. Butz, 510 F.2d 743, 753 (D.C. Cir. 1975).
46. Int'l Tel. & Tel. Corp. v. Gen. Tel. & Elec. Corp., 518 F.2d 913, 917-918 (10th Cir. 1975).
47. E.g., Morton v. Ruiz, 415 U.S. 199, 231 (1974).
48. Hutt, P.B. Food Drug Cosm. L.J. 1973, 28, 460.
49. 38 Fed. Reg. 19226 (July 19, 1973) (carcinogenic animal drug residues).
50. 42 Fed. Reg. 10412 (February 22, 1977) (carcinogenic animal drug residues).
51. 43 Fed. Reg. 8808 (March 3, 1978) (aflatoxin).
52. 44 Fed. Reg. 17070 (March 20, 1979) (carcinogenic animal drug residues).
53. 45 Fed. Reg. 72112 (October 31, 1980) (lead acetate).
54. 46 Fed. Reg. 15500 (March 6, 1981) (lead acetate).
55. 46 Fed. Reg. 24694 (May 1,1981) (estradiol).
56. 47 Fed. Reg. 14464 (April 2, 1982) (constituents policy).
57. 47 Fed. Reg. 14138 (April 2, 1982) (D&C Green No. 6).
58. 47 Fed. Reg. 22545 (May 25, 1982) (cinnamyl anthranilate).
59. 47 Fed. Reg. 24278 (June 4, 1982) (D&C Green No. 5).
60. 47 Fed. Reg. 57681 (December 28, 1982) (D&C Red Nos. 6 and 7).

RECEIVED November 4, 1983

7

Inter-Risk Comparisons

E. A. C. CROUCH and RICHARD WILSON

Energy and Environmental Policy Center, Jefferson Physical Laboratory, Harvard University, Cambridge, MA 02138

The comparison of different actions or processes for their risk content is contingent on the performance of some sort of risk assessment, which consists of the evaluation of some measure(s) of risk for those actions or processes. The particular measure(s) will depend on the reason for the assessment, for no single measure of risk is known which can encompass all aspects of risk. The need to evaluate risk measures usually requires an extrapolation of observations to new situations, a task performed by adopting models to describe how the measures vary. Such a procedure introduces various uncertainties which should be incorporated into any statements about risk. To put health risks from chemicals into perspective we compare some measures of risk for various aspects of everyday life with similar measures of risk from chemicals. In both cases we outline the models used in the risk assessment and the uncertainties in the values obtained.

The discussion of risks from any particular action, process, or system often procedes in splendid isolation, usually with protagonists and antagonists ranged on two sides of an unbridgeable gulf quoting contradictory and alarming risk estimates at each other. What we try to do in this paper is cursorily point out why the contradictions may be only apparent, the gulf bridgeable, by indicating where such apparent contradictions often arise, and then go on to help remove the isolation and alarm by providing a few examples from everyday life with which to provide comparisons. For it is often the isolation of risk estimates that make them seem alarming.

0097-6156/84/0239-0097$06.00/0

Comparisons of risks requires the evaluation in various cases of similar measures of risk, a task that requires the model ling of risky actions or processes in order to extrapolate to new situations. We will describe some of the problems of doing this, together with some of the procedures used in this paper and the approximations involved. Against the background provided by everyday risks, and with an appreciation of the approximations and uncertainties involved, we can extend our "risk list" (at least partially) to some of the products of the chemical industry and sketch a procedure for dealing with the risks which arise.

Measures of Risk

In making comparisons between risky actions or processes, as in making any comparisons, it is desirable to avoid attempting to compare unlike quantities. A quantitative risk assessment ends up with some number or range of numbers which describe(s) risk. A closer look will reveal the possibility of finding a whole set of such numbers, each of which describes some particular aspect of the risk. Each member of such a set is just one measure of that risk, and must be so treated. Comparison with a different measure of some other risk may be misleading--indeed comparison with a different measure of the same risk can be confusing. Figures 1 and 2 show how two different measures of risk of accidental death for the U.S. coal industry varied over the 20 year period from 1950 to 1970. One figure seems to indicate that the industry got substantially "safer" over that period, while an opposite conclusion may be inferred from the other. Each measure represents a different aspect of the risk of accidental death, and whether they support or deny any conclusions as to the safety of the coal industry depends, inter alia, upon a definition of "safety" in this context.

Similar apparently contradictory measures of risk may be constructed in other cases, and they are useful for emphasizing the necessity of clear definition. The purpose of the risk assessment has to be well-defined before suitable risk measures can be constructed, and comparisons between different risk measures can easily be ambiguous.

System Boundaries

Another "apples and oranges" comparison can arise when risk assessments have been performed on two or more putative alternatives, for example construction and operation of "conventional" versus "renewable energy" electric power plants. Even if the same risk measure is used in each case, little is gained in their comparison if the alternatives are not equivalent (or different) in some well-defined way. Changing the definition of equivalence may alter any conclusions to be drawn from a risk assessment.

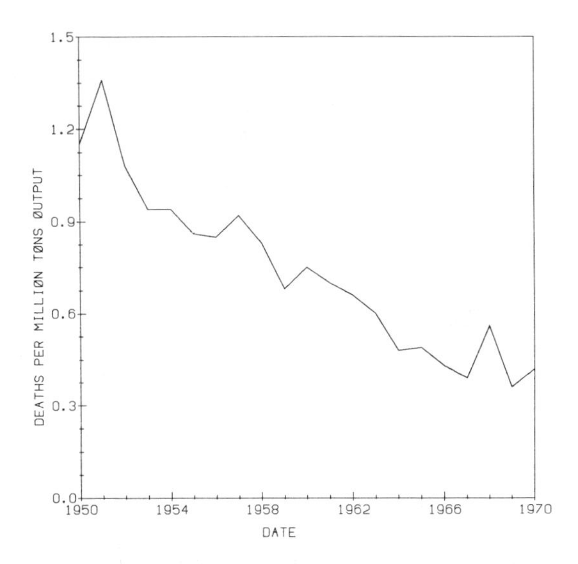

Figure 1. U.S. Coal Industry, 1950-1970. Deaths per million tons output.

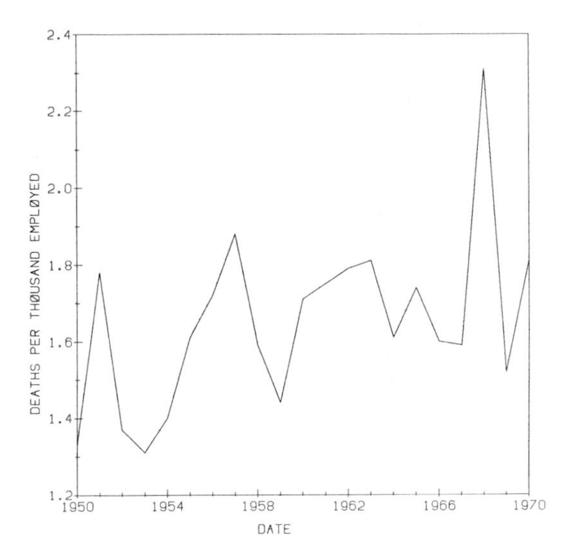

Figure 2. U.S. Coal Industry, 1950-1970. Deaths per thousand employed.

This is apparent in the example quoted, where the different types of power plants may be constructed to be equivalent in different ways. Conventional plants are usually designed to meet specific availability (probability of being able to supply power when called upon to do so) and power output goals, whereas a more sensible way of designing "renewable energy" plants, especially those powered by wind or sun which provide energy at times not dictated by man, might be for maximum energy output with little account taken of availability.

Once again the reason for the risk assessment is of paramount importance in deciding what measures of risk to compare and for which systems those measures of risk have to be evaluated.

Modelling and Uncertainty

Any risk assessment requires the implicit or explicit use of modelling to describe the process which is being assessed and associate risk with it. The model is then used to extrapolate to the situation of interest in the risk assessment. A careful consideration of even the simplest assessments will show this general pattern and assist in indicating the uncertainties which necessarily arise as a result. As an example, we take the risk of death in auto accidents.

First we need to select some useful measure of risk. For this example we will use the United States population average annual probability of dying as our risk measure. There are abundant data for the past behaviour of this measure, some of which are plotted in Figure 3. It appears that this risk has been fairly constant in the past few years, with occasional jumps such as that in 1973-1974, but no significant long term trend is apparent. On this basis we might propose that this measure of risk is a constant, with random annual variations. Such a proposal would then constitute our model, which we would fit to the data and find that the risk is, on average, 24 per 100,000 per year, with a random year to year scatter of about 10%. On this basis we might then suggest that in future this same measure of risk would also be 24 per 100,000 per year plus or minus ten per cent.

The procedure used here was to propose a plausible ad hoc model, obtain the parameters of that model by fitting to historical data, and then extrapolate (to the future) using the model. The two parameters obtained were the average value (24 per 100,000 per year) and the average annual variability (10%). There are two sources of uncertainty in this procedures, the first easy to handle but the other very difficult. The first arises in fitting the model to available data in order to estimate the model parameters. The values obtained for the parameters will be subject to the usual statistical uncertainties associated with fitting theoretical models to observed points, but such uncertainties can themselves be estimated and dealt with by standard procedures.

The second source of uncertainty is in the choice of model and the validity of the extrapolation process. Although in this case the model we chose is plausible when looking at Figure 3, a little reflection will show that it could be completely wrong--the constancy of this measure is certainly not fundamental and may have arisen in the past fortuitously. In other words, while our model may adequately fit (or describe) the data in the past, it does not follow that it gives a mechanistic description of what actually happened then nor of what may happen in the future. Alternative models can be postulated which may be better representative of the world, yet behave completely differently when extrapolated into the future. Figure 4 shows another measure of risk for auto accidents, the average number of deaths per vehicle mile travelled, which shows a declining trend with time, which trend may continue into the future. If we are interested in estimating the average risk of death in the population, using this last measure also requires an estimate of vehicle miles travelled. Extrapolating to new situations (e.g. the future or to a different country) may be more satisfactory using models of Figure 4 together with models of how vehicle miles travelled will vary, rather than using a simple fit to Figure 3, since such an approach automatically contains the intuitively obvious--that gross variations in the total amount of driving will have some effect on the numbers killed.

There is no way in which the sizes of the uncertainties introduced by failure to choose the "right" model may be rigorously estimated. To use any model one usually has to make a large number of implicit or explicit assumptions, many of which cannot be tested with available data, although obviously it helps if any models chosen agree with what data is available. Extrapolations based on models thus have to be made on the basis of plausibility, and uncertainties due to incorrect choice of model can only be guessed at if one is prepared to accept some assumptions--such as by accepting that a certain class of models encompasses the only possibilities, and finding the spread in extrapolated results for every member of that class.

Everyday Risks of Life

Bearing in mind the dangers of misinterpretation and the likelihood of errors which we have just discussed, it is useful to appreciate the magnitudes of some of the risks we face in everyday life. Table 1 presents a few such values for occupational risks of death in U.S. industries. These values will provide a useful anchoring point for comparison with some of the values we obtain later for other risks. Notice that a risk of one in a

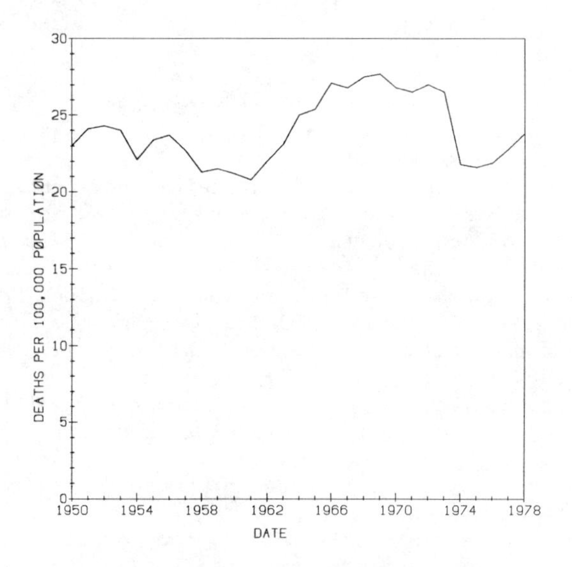

Figure 3. U.S. Motor-vehicle accident deaths, 1950-1970. Deaths per 100,000 population.

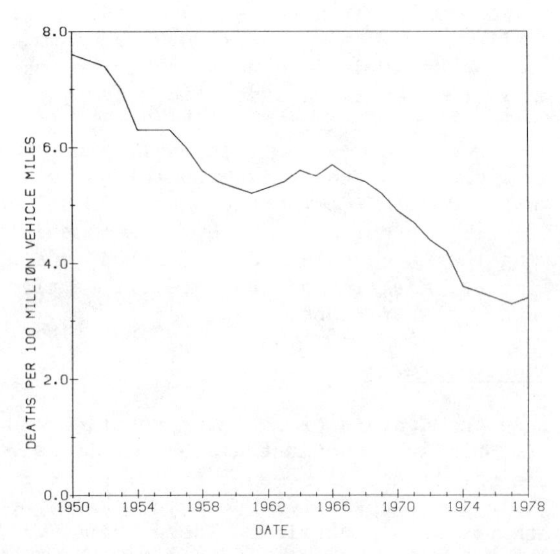

Figure 4. U.S. Motor-vehicle accident deaths, 1950-1970. Deaths per 100 million vehicle miles.

million per year applied to the whole U.S. population would result in an annual death toll of about 240, but the risks shown in Table 1 are applicable only to various subpopulations covered by the designated industry group.

Table I. U.S. Occupational Risks in 1978 or in the Year Shown

Industry Group	Annual Occupational Risk of Death (1978)	Variability (percent)	Trend
Trade	5.3×10^{-5}	15	Yes
Manufacturing	8.2×10^{-5}	8	Yes
Service and Government	1.0×10^{-4}	8	Yes
Transport and Public Utilities	3.7×10^{-4}	16	No
Agriculture	6.0×10^{-4}	9	No
Construction	6.1×10^{-4}	6	Yes
Mining and Quarrying	9.5×10^{-4}	22	No
More finely divided grouping:			
Farming	3.6×10^{-4} (1977)	7	
Stone quarries and mills	5.9×10^{-4} (1971)	20	
Police Officers (in line of duty)	2.2×10^{-4} (1978)	19	
Railroad Employee	2.4×10^{-4} (1977)	7	
Steelworker (accident only)	2.8×10^{-4} (1972)	?	
Firefighter	8.0×10^{-4} (1972)	?	

The values obtained in this table were obtained by applying a particular model. For each year from 1955 to 1978 the measure shown was computed by dividing reported occupational deaths in the industries listed by the reported average work force. It was then assumed that this measure varies linearly with date (and is independent of variations in occupational populations, average hours worked, average experience of the working population, etc., except insofar as these things vary linearly with date), so that a simple linear time trend could be extracted from the raw data. The value of the resultant fitted model in 1978 is listed in Table I if there was a significant time trend, otherwise the average value over 1955 to 1978 is listed. The variability recorded represents the standard deviation of the observed values about the theoretical model, as a percentage of the 1978 estimate. For reporting on historic data the model adopted is often adequate, but for extrapolation purposes it may be gravely inadequate--examples of extrapolation hwere would be for prediction of future risks, prediction of risks in individual industries (components of the industrial groupings listed), predictions of risks in similar industry groups in other countries.

There is clearly considerable variation between industrial groupings (by a factor of ten) in this measure of the occupational risk of death borne by employees. The variation is probably larger between individual industries, since the values given are averages over industry groups which often contain substantially different components. Some idea of this variation may be seen from the second part of Table 1, which shows similarly computed measures of risk for a few subgroups of employees. Comparison of the two tables shows variations up to a factor of ten within the groupings of Table I, and it is almost certain that larger variation could be found with further studies of other subgroups. Nevertheless a general conclusion is that occupational risks of death lie in the range of one in ten thousand to one in a thousand per year, with the ordering of industries being approximately as one would expect.

Table II lists a set of commonplace risks of accidental death in the United States. As in Table I, significant time trends have been factored out of these values--they represent a value estimated for 1977 and 1978, based on a sequence of several years. For comparison with these accidental risk rates, the risk of death by homicide in the U.S. in 1976 was about 9 per 100,000. Risks of accidental death in various sports is shown in Table III. Perhaps these could be interpreted as showing what we are prepared to do to ourselves, compared with what we are prepared to have imposed upon us. These values correspond to the annual average risk of death for those participating in the sport. There is a large uncertainty in the value of most of these risks, corresponding to a factor of 2 or 3, since although the number of deaths is usually accurately known, the number of people participating in each sport is highly uncertain. From this list it would appear that going up into the air in almost any (noncommercial) way or down into the deapths of the sea (scuba diving) in any way are both associated with comparatively high risks!

One might immediately wish to start comparing the values in the various tables, and this is possible. But recall the earlier discussion on comparisons of different things. Although the risk measures are similar in the various tables, the purpose of any such comparison must be made clear before attempting it, for a different risk measure might be more appropriate. For example most employees are at work for a large fraction of the time throughout the year, whereas sports are played only intermittently. A measure of risk which took this disparity into account may be of greater value.

Table II. Accidental Risks of Death in the U.S. in 1977 or 1978

Accident	Annual Average Risk of Death	Variability	Trend
Motor Vehicle	2.4×10^{-4}	10	No
All Home Accidents	1.1×10^{-4}	5	Yes
Fall	6.2×10^{-5}	6	Yes
Drowning	3.6×10^{-5}	7	No
Fire	2.8×10^{-5}	5	Yes
Inhalation/Ingestion of Objects	1.5×10^{-5}	8	No
Accidental Poisoning	1.4×10^{-5}	<10	No
Firearms (accidents)	1.0×10^{-5}	8	No
Electrocution	5.3×10^{-6}	5	No
Tornado	6×10^{-7}	100	No
Flood	6×10^{-7}	100	No
Lightning	5×10^{-7}	18	No
Tropical Cyclone/ Hurricane	3×10^{-7}	160	No
Bite/Sting	2×10^{-7}	13	No

Table III. Accidental Risks of Death in U.S. Sports--Averaged Over Several Years. Upper Limits are Given Where The Risk is Based on 4 or Fewer Deaths, by Assuming that Deadly Accidents Are A Poisson Process with Total Number Proportional to Person-Years at Risk. Uncertainty is a Factor of 2 to 3.

Sport	Annual Average Risk of Death
Professional Stunting	$< 1 \times 10^{-2}$
Air Show/Air Racing and Acrobatics	5×10^{-3}
Flying Amateur/Home Built Aircraft	3×10^{-3}
Sport Parachuting	2×10^{-3}
Professional Aerial Acrobatics	$< 2 \times 10^{-3}$
Hang Gliding	8×10^{-4}
Mountaineering	6×10^{-4}
Glider Flying	4×10^{-4}
Scuba Diving	4×10^{-4}
Spelunking	$< 1 \times 10^{-4}$
Boating	5×10^{-5}
College Football	3×10^{-5}
Hunting	3×10^{-5}
Swimming	3×10^{-5}
Ski Racing	2×10^{-5}

Cancer Risks

We have been interested for some years in the cancer risks posed by contaminants and additives in food, water and air, and especially in how to make estimates of those risks. To give some anchoring points in discussing such risks, Table IV presents approximate values for the U.S. population lifetime probability of dying from various cancers. These values are obtained simply by finding the proportion of all deaths which are due to a given tumor and they are approximate in that this procedure only gives an approximation to the stated risk. A risk measure similar to that shown in previous tables is obtained by dividing these values by the average lifetime expressed in years (about 72), to give the second column in the table. The total cancer risk shown here is comparatively large, and it is interesting to attempt to partially explain its origin, since some part is presumably due to exposure to environmental agents including materials in food, air and water. One such environmental agent is radiation, so Table V attempts to show the effects of the major population exposures. Of course we have used a model to estimate these risks. The model assumes that excess risk of cancer (not distinguished from death here, since the difference is smaller than the uncertainties) is proportional to radiation dose, with a constant of proportionality corresponding to 1 cancer death per 5000 man-rem of radiation exposure. The latest best estimates of the effect of ionizing radiation exposure by the Committee on the Biological Effects of Ionizing Radiation would give slightly lower values than those shown.

The models used in estimating risks from exposure to other environmental agents are usually very similar to that used for radiation exposure, although there is substantially more data, especially human data, in the case of radiation. For exposure to chemicals, we assume a model which states that at low doses the lifetime probability (R) of a cancer is proportional to the lifetime average dose rate (d) when the latter is measured as a fraction of bodyweight consumed per day (usually expressed in milligrams per kilogram per day). This model is:

$$R = 1 - (1 - \alpha) * \exp(-\beta d/[1 - \alpha])$$

$$\sim \alpha + \beta d \text{ (at low doses)}$$

where α is the probability of cancer in the absence of the material. The constant of proportionality (β) is called the potency of the material, and it can be measured in experimental animal studies in which animals are exposed to the material for a lifetime. From many such studies, we have found that the potency of a particular material in one species can be used to estimate that material's potency in another species with an

Table IV. Lifetime Risk and Annual Average Risk of Death from Cancer in the U.S.*

Type	Lifetime Risk	Average Annual Risk
All Cancers	0.20	2.8×10^{-3}
Buccal cavity, pharynx, respiratory	0.050	7.2×10^{-4}
Digestive organs and peritoneum	0.053	7.5×10^{-4}
Bone, connective tissue, skin breast	0.022	3.1×10^{-4}
Genital organs	0.022	3.2×10^{-4}
Urinary tract	0.008	1.2×10^{-4}
Leukemia, other blood and lymph	0.018	2.6×10^{-4}
Other	0.019	2.7×10^{-4}

*The uncertainty in all these values is about 20%.

Table V. Cancer Risks from Radiation Exposures

Type	Average Annual Risk
Natural background (average U.S., sea level)	2×10^{-5}
U.S. average medical diagnostic x-ray	2×10^{-5}
Excess due to living in masonry building rather than wood	5×10^{-6}
Cosmic Rays:	
Airline pilot (50 hrs./mo. at 12 km altitude)	4×10^{-5}
One transcontinental round trip by air per year	1×10^{-6}
Frequent airline passenger (4 hrs./wk.)	1×10^{-5}
Living in Colorado compared with New York	8×10^{-6}
Camping at 15,000 ft. for 4 mos./yr.	2×10^{-6}

uncertainty corresponding to a factor of about 5, and so similarly we expect that the potency of a material in a laboratory animal can be used to predict its potency in humans to within a similar factor. Using such models allows us to construct Table VI, showing some of the annual average risks (lifetime risk divided by 72 years average lifetime) from carcinogens in food and drinks, risks from smoking (based on human data) and a risk of cancer from polycyclic organics in average U.S. city air pollution.

Table VI. Some Everyday Cancer Risks from Common Carcinogens

Action	Average Annual Risk	Uncertainty
One 12.5 oz. diet soda daily (saccharin)	1×10^{-5}	factor of ~10
Average personal saccharin consumption	2×10^{-6}	
4 tbsp peanut butter per day (aflatoxins)	8×10^{-6}	
One pint of milk per day (aflatoxins)	2×10^{-6}	
Miami/New Orleans drinking water	1×10^{-6}	
1/2 lb charcoal broiled steak/week (cancer only; heart attack etc., extra)	3×10^{-7}	
Average smoker (cancer only)	1.2×10^{-3}	factor of 3 (human data)
(all effects)	3×10^{-3}	
Person sharing room with smoker	1×10^{-5}	factor of ~10
Air pollution (polycyclic organics)	1.5×10^{-5}	

One interesting consequence of using a linear model for relating dose and effect is that the total number of cancers caused is independent of how the dose is spread out in the population, provided the risk to each individual remains small (less than about 10% lifetime risk, or 0.0014 annual risk). Thus we can use the same simplified formula:

$$R = \beta d$$

to estimate lifetime risk (R) to an individual exposed to a lifetime average dose rate d, or to estimate a number of cancers (R) in the lifetime of a population exposed to a total population lifetime average dose rate d. Introducing two other factors, an interspecies factor K to convert from potency as measured in animals to potency in humans, and a dispersion factor I which is the proportion of total production of a chemical which is finally absorbed by humans, allows us to write the expected annual cancer deaths (n) due to an annual production P of chemical as:

$$n = \beta KIP \times 0.254$$

where
- n = Number of cancers expected per year
- β = Potency in test animal (in kg-day/mg)
- K = Interspecies factor
- I = Fraction of production absorbed by humans
- P = Production of chemical (in lbs/year)
- 0.254: Factor to convert between units and to convert from lifetime risk to annual risk

Our best estimate of the factor K is that it varies randomly from chemical to chemical, with log (K) following a normal distribution with mean value zero and standard deviation 0.65 for comparisons between rats or mice and humans--corresponding to the factor of about 5 mentioned above. If desired, an extra multiplicative factor can be introduced into the equation above to account for the possibility that the dose response curve may not be linear at low doses. The most likely value of such a factor is less than one, but we shall assume that it is equal to one.

This formulation allows the construction of a very preliminary set of relative risk estimates, the initial part of which is shown in Table VII where a hazard index is contructed. The factor K is omitted, since it is common to all the entries, and we only enter a single value for the risk measure rather than a probability distribution which would be preferable, but this is only intended as a crude first approximation to locate possible danger signs. The calculations are performed sequentially by first computing what would be the consequences of no dispersion, giving a hazard index comparable to those used in discussions of radioactive materials. In several cases no single point estimate is possible, because the animal test gave a result which was not statistically significant, but we can then give an upper bound based on the sensitivity of the animal test. At this stage any estimate that we can make of the dispersion factor would be not much more than guesswork, although modelling has been performed for specific plants producing specific chemicals, so we invite readers to supply their own guesses or computational results.

Despite the crudity of the approach suggested, Table VII does provide a useful initial exercise in risk assessment, and will probably draw attention to those chemicals needing greater study--those which are capable of causing relatively large numbers of cancers. In particular it graphically demonstrates the need for some estimation of dispersion factors, and also the relative insensitivity of carcinogen testing methods.

Conclusions

Given the uncertainties inherent in any risk assessment and the ambiguities in the results discussed in the first part of this paper, some might think that there is little point in even attempting to compare risks quantitatively. Our own view is that because of such uncertainties and ambiguities, quantitative risk assessment is essential, mainly because it forces the analyst to define where the ambiguities lie and to quantify the uncertainty, and makes explicit those areas where opinions or judgement are necessary. Any good attempt at risk assessment will result in a much clearer view of the problem under consideration, with quantitative statements of what is known and what is not known, together with statements of the uncertainties in both the known and unknown parts.

The tables of risks shown here should not be considered the outcome of any detailed risk assessments, but should be taken as rough guides to indicate the magnitudes of commonly faced risks. Perhaps someone will be provoked into deeper analysis of some of them, an examination which might lead to their reduction. The everyday risks from some carcinogens were estimated using simple models. Using the same simple models allows an initial estimate of the hazards associated with production of chemicals, although further analysis is clearly needed to define the dispersion of those chemicals to the population. Nevertheless, the process provides an initial ordering of chemicals and clearly identifies those for which further risk analysis is essential. We would like to see the chemical industry performing similar analyses on all their products (and also on their effluents) as a first step in deciding where to expend resources on risk reduction.

Table VII. Construction of a Hazard Index for Chemicals. The Production Figures Correspond to the Year Indicated

Material	Potency in mouse (kg-day/mg)	Production (million lbs/year)	Hazard Index (deaths per yr)
Insecticides/pesticides/ herbicides			
Malathion	1.0×10^{-4}	16 (76)	400
Trifluralin	1.1×10^{-3}	11.4 (77)	3000
Methoxyclor	$< 7.5 \times 10^{-4}$	5.5 (76)	< 1000
Pentachloronitrobenzene	1.4×10^{-4}	3 (71)	110
Endosulfan	$< 1.8 \times 10^{-1}$	1 (74)	< 50,000
Chloropicrin	$< 4.4 \times 10^{-3}$	4.8 (76)	< 5000
Azinphosmethyl	2.1×10^{-2}	3.1 (76)	17,000
1,2-Dibromoethane	4.8×10^{-2}	1.2 (74)	15,000
Dicofol	1.9×10^{-2}	4 (77)	19,000
Coumaphos	4.7×10^{-2}	0.4 (76)	5000

Table VII. Continued.

Material	Potency in mouse (kg-day/mg)	Production (million lbs/year)	Hazard Index (deaths per yr)
Insecticides/pesticides/ herbicides			
Fenthion	8.5×10^{-2}	0.2 (76)	4000
Anilazine	$< 1.9 \times 10^{-3}$	0.2 (76)	<100
Piperonyl Butoxide	1.5×10^{-4}	1 (76)	40
Aldicarb	$< 5.1 \times 10^{-1}$	1.6 (76)	<200,000
Diazinon	2.6×10^{-3}	7 (76)	5000
p,p'-ethyl-DDD	3.0×10^{-4}	0.2 (76)	15
Methyl Parathion	$< 2.9 \times 10^{-2}$	53 (75)	<400,000
Miscellaneous products			
1,2-Dichloroethane	2.1×10^{-3}	11,110 (80)	6,000,000
Allyl Chloride	$< 1.8 \times 10^{-3}$	295 (75)	<130,000
1,1,2-Trichloroethane	5.9×10^{-3}	124 (75)	190,000
Titanium Dioxide	$< 4.2 \times 10^{-5}$	1.9 (77)	<20
Dimethyl Teraphthallate	4.3×10^{-4}	2800 (77)	300,000
Ethyl Tellurac	$< 1.4 \times 10^{-3}$	6 (79)	2000
Phthalic anhydride	$< 3.0 \times 10^{-5}$	1000 (79)	< 8000
N-Nitrosodiphenylamine	7.4×10^{-5}	1.3 (77)	20
Tetraethylthiuram disulfide	2.0×10^{-3}	1.2 (77)	600
Phenol	$< 2.0 \times 10^{-4}$	2400 (79)	<120,000
4,4'-oxydianiline	4.8×10^{-3}	<1 (74)	<1200
Trichloroethylene	7.3×10^{-4}	267 (80)	50,000
1,1,1-Trichloroethane	1.7×10^{-5}	650 (80)	3000
Nitrilotriacetic acid	3.8×10^{-4}	150 (70)	14,000
Tetrachloroethylene	1.6×10^{-3}	734 (76)	300,000
1,1,2,2-Tetrachloroethane	1.2×10^{-2}	40 (75)	120,000
1,1-Dichloroethane	1.0×10^{-4}	11,794 (80)	300,000
DL-menthol	$< 4.8 \times 10^{-4}$	146 (77)	<18,000
Chemical agents and reagents			
2,4-Dinitrotoluene	$< 2.4 \times 10^{-3}$	272 (77)	<170,000
Tris	5.9×10^{-3}	3 (76)	4000
1,4-Dioxane	1.0×10^{-3}	18 (76)	5000
1,2-Dibromoethane	4.8×10^{-2}	290 (80)	4,000,000
2,4,6-Trichlorophenol	9.5×10^{-4}	.002 (77)	0.5

Continued on next page

Table VII. Continued.

Material	Potency in mouse (kg-day/mg)	Production (million lbs/year)	Hazard Index (deaths per yr)
Dyes, pigments and intermediates			
5-nitro-o-toluidine	2.0×10^{-3}	0.13 (77)	70
Aniline hydrochloride	$< 2.0 \times 10^{-4}$	< 400 (77)	< 20,000
C.I. Vat yellow 4	5.0×10^{-5}	0.15 (80)	2
1-Phenyl-3-methyl-5-pyrazolone	$< 1.0 \times 10^{-4}$	0.02 (77)	< 0.5
p-Cresidine	1.0×10^{-2}	0.6 (77)	1500
2,4-Diaminotoluene	2.1×10^{-2}	233 (77)	1,200,000
Diarylanilide yellow	$< 2.7 \times 10^{-5}$	13 (79)	90

Acknowledgments

Our work on risk assessment, more fully described in our book Risk Benefit Analysis (Ballinger, 1982), would not have been possible without financial support from the Electric Power Research Institute (Technical Agreement TSA 79-294), the General Electric Foundation, the Cabot Corporation, Monsanto and Dow Chemical. Subsequent work on carcinogenic risk assessment has been funded by the U.S. Department of Energy under Contract No. DE-AC02-81EV10598.

RECEIVED November 4, 1983

8

Uncertainty and Quantitative Assessment in Risk Management

M. GRANGER MORGAN

Department of Engineering and Public Policy, Carnegie-Mellon University, Pittsburgh, PA 15213

Quantitative assessment is required as a tool in risk management because many risk processes are too complex to be understood without such assessment and because psychological heuristics introduce bias in many judgmental estimates of risk. Uncertainty can enter risk assessment through exposure processes and/or through effects processes and may involve the value of coefficients or the actual functional relationship among variables. Some characteristics of a "good" quantitative assessment are enumerated and a number of reasons for explicitly incorporating uncertainty in quantitative assessment are advanced. The analytical implications of alternative levels of uncertainty are discussed. A general software system for the support of such analysis is described and brief examples are provided. For at least some risk assessment/risk management problems the development of appropriate quantitative analytical tools may be able to provide contesting parties with a common framework within which to address the problem and argue their respective views in a more systematic and somewhat less adversarial fashion.

My thesis in this paper is that in order to be "good", a quantitative risk assessment must characterize and deal with the major uncertainties associated with the problem. But before I can address this issue I must first concern myself with what I mean by"good." Good against what criteria? Good given what objectives?

People undertake quantitative risk assessments, and other quantitative policy assessments for a variety of reasons. These reasons include:

0097-6156/84/0239-0113$06.00/0

1. To get the answer to a specifically formulated policy question.
2. To illuminate and provide insight on a set of policy issues.
3. To provide substantiation and arguments to support your views in an adversarial procedure.
4. To persuade others that you have got things under control, know what you are doing, and should be trusted. Performing policy analysis can sometimes do this because it draws on the tools and images of science and social science and the paradigm of "rational decision-making".
5. Because the law says you must.
6. Because other people expect you to.
7. Because it is not clear what else to do and the situation is such that you feel you must do something.
8. Because doing quantitative policy analysis can be fun and professionally rewarding.

There is a clear difference in kind between the first three of these reasons, the next four, and the last. The first three are substance-focused because the specific substance and output of the analysis is of primary importance. The next four are process-focused because the process of _doing_ quantitative analysis may, in these cases, be more important than the specific details or findings involved. The eighth and final reason is analyst-focused. In my judgment, many of the best quantitative policy analysts I know are in the business for this reason. Of course, if you ask them why they are in the business they will generally give you answers that sound like one or more of the substance-focused reasons that I have identified.

Many people trained in the paradigms of science and engineering are likely to find it hard to take process-focused reasons for doing quantitative policy analysis very seriously. These reasons are, however, very real, and based at least on my own experience are often more important in promoting the use of quantitative risk assessment and other policy analysis than are substance-focused reasons.

While process-focused reasons are perhaps the most common reason for engaging in quantitative assessment, they can't be used as publicly stated justifications for doing analysis. Take reason 4 as an example. Performing quantitative assessment can be a good way to persuade others that I know what I'm doing because they assume I'm doing analysis to get answers to policy questions or to better illuminate the policy issues. Since the policy making process is often too convoluted for outside observers to follow, people's perception of me may be quite uneffected by the fact that none of the quantitative analysis that I perform actually has any impact on, or significance for, the policy decisions I make. But, if I publicly admit this fact, then, of course, the analysis is no longer of value to me. I must maintain the fiction that my motivation is one of the first two substance-focused reasons.

In a similar way, the third of the substance-focused reasons relies for its success on an assumption among most participants and observers that one of the first two substance-focused reasons is actually operating. If I can easily demonstrate, for example, that the inputs to the analysis were artfully chosen to get the answer desired, the effectiveness of the analysis as an adversarial tool is greatly diminished.

Thus, we come to a very interesting conclusion. While we have been able to list at least eight reasons why organizations and people commission and perform quantitative assessments, only the first two of these reasons are relevant when we ask the question what are the characteristics of "good" quantitative assessment?

The first of these two reasons (to get answers to a specifically formulated policy question) is the one advanced by classical decision analysis. Ocassionally one actually does encounter a problem in which a good piece of analysis provides an answer that can be directly implemented without further consideration. However, such situations are so rare in practice that for the balance of this discussion I will focus exclusively on my second reason ...namely, quantitative risk and other policy assessments are undertaken to provide insight and to inform the policy making process.

In the specific context of risk assessment, this translates to the statement that quantitative risk assessment is undertaken to provide a better understanding of the character, magnitude and extent of specific risks as well as an understanding of how these risks compare with other existing or potential risks. Quantitative risk assessment doesn't tell us whether and how to manage a risk but it may provide insights that make it easier to select risk management strategies that are consistent with our values and beliefs. Thus, it can "inform" the policy process of risk management decision-making.

Of course, there are many policy problems to which quantitative analysis can contribute little or no insight. Hence, it is worth asking is quantitative risk analysis worth doing? Can it provide insight and inform the policy process of risk management in ways that would otherwise not be possible? For a large number of risks I think the answer is yes, for two reasons. First, the exposure and effects processes involved in many risks are too complex for people to fully understand without the assistance of quantitative models. Second, there is growing experimental evidence (1,2) that the psychological heuristics that people use in making judgments about risks do not work very well in the context of many risks, particularly technologically-based risks to human health, safety, and the environment. Quantitative risk assessment can provide a vehicle for avoiding at least some of the biases that are introduced by the operation of these heuristics.

Why is it "Good" to Characterize and Deal With Uncertainty in Risk Assessment and in Other Quantitative Policy Analysis?

Some years ago I wrote an editorial in Science (3) that began to list what I think are the characteristics of "good" policy analysis. Since then I have thought quite a bit more about this question. If the objective of good policy analysis is to inform the policy making process, then my current list of the attributes of "good" quantitative policy analysis reads as follows:

- Clearly define the boundaries of the analysis and provide a careful rationale for their selection.
- Develop the models (or analytical tools) used in the assessment in an iterative fashion so as to assure, through careful sensitivity analysis, that all important variables are identified and properly incorporated and that the models chosen are as robust as possible. Explicitly describe this process and use the findings to justify the choice of models.
- Characterize and deal with all scientific or technological uncertainties as completely and as explicitly as possible.
- Explicitly identify all value assumptions and, to the extent possible, treat them in a parametric fashion so that a variety of people with different value orientations can use the analysis to reach their own conclusions.
- Present results in a clear open manner which makes all assumptions and operations explicit and allows others to easily verify, use, modify and extend the analysis.
- To the extent possible, describe how the conclusions reached may be effected by the problem boundaries that were selected.

I could go through each of these six points and develop supporting arguments, but because the subject of this paper is the importance of adequately characterizing and dealing with uncertainty let me focus just on this one attribute on my list.

The standard decision analysis literature (4,5) identifies a number of circumstances in which it may be important to explicitly characterize and deal with uncertainty when performing analysis. These include:

- When you are performing an analysis in which peoples' attitude toward risk is likely to be important (e.g. when people are risk averse).
- When you are performing an analysis in which uncertain information from different sources must be combined (e.g. when the mean of the output estimate cannot be obtained simply by operating on the means of all the input estimates).
- When a decision is to be made about whether to expend resources to acquire additional information (e.g. problems involving the value of information).

In a Ph.D. thesis recently completed under my supervision, my colleague, Max Henrion (6), has explored a fourth reason supported by the axioms of conventional decision analysis:

- When the losses go as the third or some higher order of the decision variable.

He defines the EVIU or expected value of including uncertainty and compares it with other standard measures such as the EVPI for a variety of examples.

However, the paradigm and formalism of decision analysis doesn't capture everything that is important in real policy analysis environments. There are a number of other reasons why it may be important to characterize and deal with uncertainty in analysis even when the EVIU in a decision analytic context is zero or very small. Here are several:

- There is considerable empirical evidence to suggest that due to a variety of heuristics employed in human thought processes cognitive biases may result in "best estimates" that are not actually very good. Even if all that is needed is a "best estimate" answer the quality of that answer may be improved by an analysis that incorporates and deals with the full uncertainty.
- Model building is necessarily an iterative process and to some extent an art form. The search for an adequate and robust model to handle the problem at hand may proceed more effectively and to a surer conclusion if the associated uncertainty is explicitly included and can be used as a guide in the process of model building and refinement.
- In the real world, a decision is rarely made on the basis of a single piece of analysis (though this is, of course, the implicit assumption of most decision analysis). Further it is rare for there to be one discrete decision. Rather a process of multiple decisions spread out over time is the more common occurrence. Any given piece of analysis is likely to be more useful in such an environment if it characterizes the associated uncertainty in a fairly complete fashion, thus allowing the user(s) to better evaluate it in the context of the various other factors being considered.
- Many problems in technology and public policy involve complex mixtures of disagreements over issues of value and of fact. Analytical procedures which explicitly characterize and deal with technical uncertainty (and parameterize or otherwise deal explicitly with issues of value) can help to produce a clearer separation of the two. This may help to lead to a more open and rational decision process.
- Policy analysts have a professional and ethical responsibility to present not just "answers" but also a clear and explicit statement of the implications and limitations of

their work. Analyze the attempts to fully characterize and deal with important associated uncertainties helps them to better execute this responsibility.

Sources of Uncertainty in Quantitative Risk Assessment

Figure 1 lays out a framework that I have found very useful in thinking about technologically-based risk to health, safety and the environment (7). In this formulation various events or activities set in motion exposure processes that expose objects and processes in the natural and human environment to the possibility of change. Effects processes then occur (either sequentially or concurrently) that give rise to changes or effects. People look at these effects and perceive something. These perceptions of what actually occurred are then valued, some as good, some as bad.

In my mind, the problem of performing quantitative risk assessment translates to the problem of building good quantitative descriptions or models of the effects and exposure processes involved in the particular risk problem of interest. While very important in the policy process of risk management, the human perception and valuation processes illustrated on the right hand side of Figure 1 are not part of quantitative risk assessments as it is usually defined and practiced.

Since risk is "the chance of injury or loss" risk assessment necessarily must deal with uncertainty. Uncertainty may enter through exposure processes, through effects processes or through both. This uncertainty may take at least three forms:

1. The values of all the important variables involved are not or cannot be known, and precise projections cannot be made.
2. The physics, chemistry, and biology of the processes involved are not fully understood, and no one knows how to build precise predictive models.
3. The processes involved are inherently probabilistic, or at least so complex that it is infeasible to construct and solve precise predictive models.

As discussed in the next section, there is an important difference between the first of these that involves uncertainty in the variables or coefficients of a model and the second which involves uncertainty in the functional form of the model itself. It is also important to differentiate these uncertainties from uncertainties in decision variables (e.g. value assumptions, policy choices, etc.). In general, I believe that such variables are best parameterized rather than treated as uncertain. The information about the uncertainty attached to risk processes usually fits into one or more of the following five categories listed in order of increasing problem uncertainty:

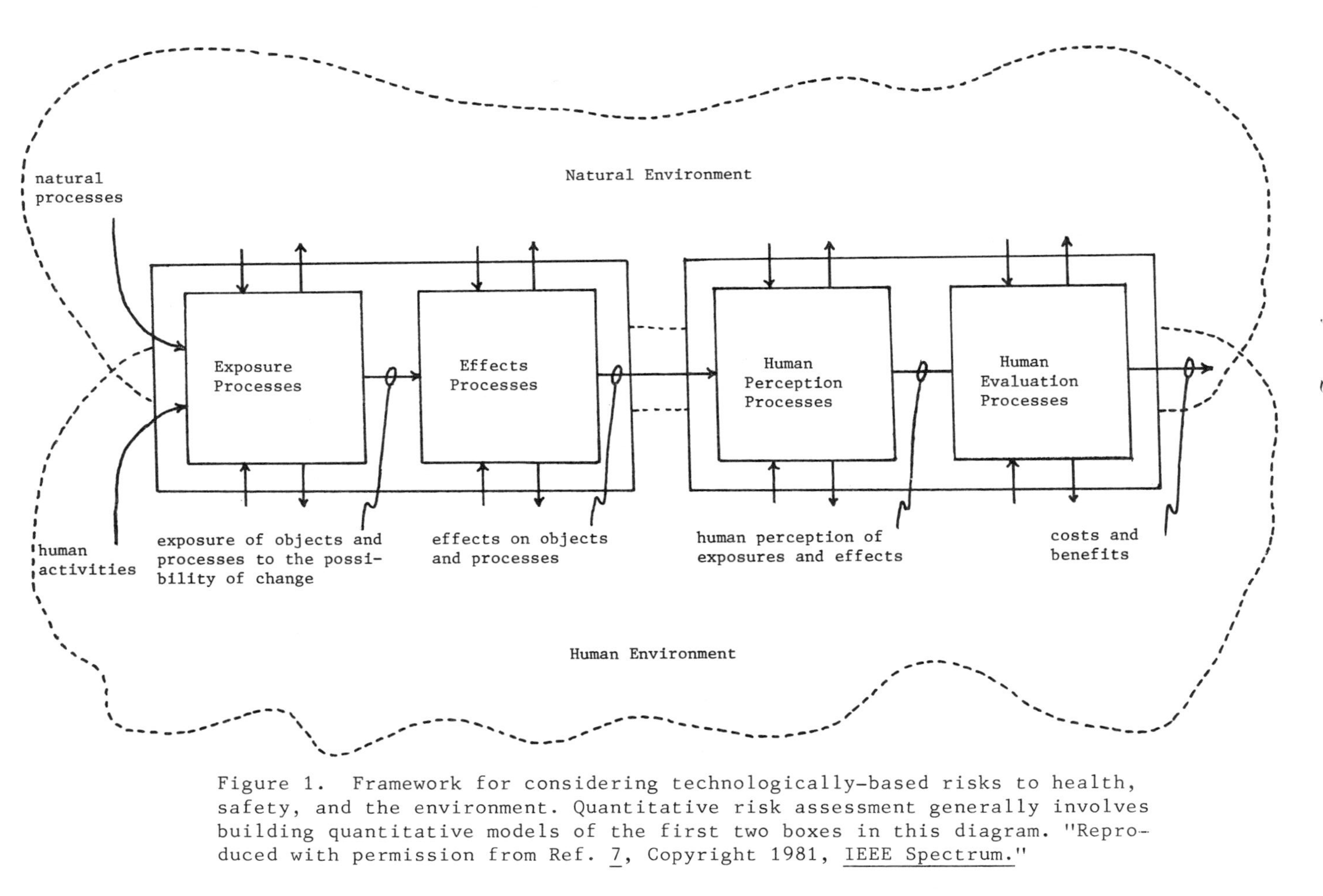

Figure 1. Framework for considering technologically-based risks to health, safety, and the environment. Quantitative risk assessment generally involves building quantitative models of the first two boxes in this diagram. "Reproduced with permission from Ref. 7, Copyright 1981, IEEE Spectrum."

1. Good direct statistical evidence on the process of interest is available. This is clearly the most desirable situation, but is rare for most categories of risk problems.
2. The process can be disaggregated with analytical tools --such as fault trees, event trees, and various stochastic models--into subprocesses, for which good statistical evidence is available. Aggregate probabilities can then be constructed.
3. No good data are available for the process under consideration, but good data are available for a similar process and these data may be adapted or extended for use either directly or as part of a disaggregated model.
4. The direct and indirect evidence that is available is poor or incomplete and it is necessary to rely to a very substantial extent on the physical intuition and subjective judgment of technical experts.
5. There is little or no available evidence, and even the experts have little basis on which to produce a subjective judgment.

Unfortunately a very substantial fraction of the quantitative risk assessment problems of concern today fall into categories 3, 4 or 5 of this classification.

Analytical Strategies and Tools for Dealing with Uncertainty

Given a risk assessment problem that involves uncertainty in the value of model coefficients there are a variety of analytical strategies which an analyst can adopt. These include:

Ia Perform single-value-best-estimate analysis and ignore the uncertainty.

Ib Perform single-value-best-estimate analysis. Then acknowledge the uncertainty, perform various sensitivity calculations, and provide a qualitative and/or quantitative discussion of the uncertainty.

IIa Estimate some coefficient of uncertainty, such as the standard deviation, for each important model coefficient and then use analytical procedures for "error propagation" to propagate this uncertainty through the analysis.

IIb Characterize uncertain coefficients as subjective probability distributions and then propagate this uncertainty through the analysis, usually through the use of stochastic simulation.

IIIa Treat some coefficients parametrically, performing the analysis for a variety of plausible values of each of these coefficients.

IIIb Perform order-of-magnitude based bounding analysis which does not produce unique "answers" but rather estimates bound on the range of possible answers.

For simplicity I will refer to the first two of these as single-value-best-estimate analysis, to the second two as probabilistic analysis and to the final two as parametric/bounding analysis. Most quantitative policy analysis, including most risk assessments, performed today use single-value-best-estimate techniques, much of it of type Ia. A fair number of analyses make some modest use of parametric/bounding techniques. Only a handful use probabilistic techniques.

Which analytic strategy or mix of strategies is appropriate for a given risk assessment problem? The answer depends largely upon the amount of associated uncertainty. I have listed the strategies roughly in order of increasing problem uncertainty. Thus, strategy Ia is most appropriate when model coefficients are really quite well known and strategy IIIb is most appropriate when very little is known about the value of these coefficients.

Having identified some available analytical strategies and arranged them roughly in order of appropriateness for increasing uncertainty the obvious next step is to provide some clearer guideline on when to move from one strategy to the next. Unfortunately, I can't do this. Within limits the choice of which analytical strategy to employ seems to me to depend largely on the taste or aesthetic judgment of the analyst. In watching myself perform assessments, I know that there are problems with so much uncertainty that I do not feel comfortable using a probabilistic approach but feel instead that a parametric/bounding approach is more appropriate. So far I haven't succeeded in putting down a coherent list of attributes that dictate my choice. I'm working on it.

The various strategies we have listed are all intended to deal with uncertainty in the value of model coefficients. There is a second kind of uncertainty that must also be considered. This is uncertainty about the correct functional relationships within the model. In general, this is a much tougher problem. Exploratory studies, changing the model around to learn what matters and what doesn't, and various order-of-magnitude kinds of arguments that do the same thing are about the best approach I can offer. Clearly an analytical environment which encourages and makes such exploratory analysis easy to do is most desirable.

There is one general conclusion I do feel quite comfortable in drawing. Far too much risk assessment that is done as single-value-best-estimate analysis should in fact be done as probabilistic analysis. There are undoubtedly several reasons for this. Performing probabilistic analysis can get analytically messy. Obtaining subjective judgmental estimates of uncertain coefficients can be awkward and is subject to a variety of pitfalls. And, while most people appear to be quite comfortable with such basic notions of uncertainty as "odds" unless care is taken, the results of probabilistic analysis can become somewhat difficult to communicate to a semi-technical or non-technical audience.

Max Henrion and I are heavily involved in research directed at reducing these obstacles, particularly the first. As part of his Ph.D. thesis, Henrion developed a software system called DEMOS which makes it roughly as easy to perform full probabilistic analysis as it is to perform conventional single-value-best-estimate analysis. Now, under NSF support, DEMOS has been re-written in clean transportable PASCAL and we are embarking on a series of studies to explore how to most effectively provide software support for probabilistic risk assessment and other probabilistic policy analysis.

While more detailed descriptions of DEMOS are available elsewhere (8,9,10,11) it seems appropriate to provide here a very simple illustration of the system. Suppose one is concerned with estimating the mortality impact of a chemical pollutant which for convenience I'll call XYZ. In the problem of interest, there are two sources of XYZ exposure. The strength of these source terms, call them S1 and S2, is uncertain. The total population at risk is P, also a somewhat uncertain quantity. The exposure process is such that exposure is proportional to source strength through a constant of proportionality, C, which is known only approximately. Finally, while a linear damage function with no threshold is known to be appropriate the slope of the damage function, D, is uncertain. A simple quantitative model of the potential health impact can thus be written:

health impact = C * (S1 + S2) * D * P

Suppose my "best-estimate" values for the coefficients are

C = 10^{-9} $\mu gm^{-3}/g$

S1 = 10 g/sec

S2 = 0.7 g/sec

D = .0025 deaths/person- $\mu g\text{-}m^{-3}/yr$

P = 1500 people

Converting the emission rates to a yearly basis and substituting we get an estimated health impact of 1.27 deaths/year. The following is a transcript of an interactive session with DEMOS that performs this same calculation. Underlined portions are what I typed (answers that appear in square brackets after questions from DEMOS are default values. If the user hits carriage return, this answer is assumed):

```
Welcome to DEMOS, version Zero.  01-June-82
Starting a new project.

Do you want to start a new project? [Yes]: Yes

Name of project? [V16]: XYZ
```

```
Description: Health impact of XYZ

Author: G. Morgan


Project XYZ is ready to be defined.
Type "Help" if needed.
>variable

Name of Variable? [v17]: EXD

Title: Health impact

Units: XD/YR

Description: Health impact of XYZ in excess deaths per year.

Definition: C*(S1+S2)*Spy*D*P

?        : Undefined variable
C is undefined.
Do you want to define it? [Yes]:

Name of Variable? [C]:

Title: Exposure coefficent

Units: ugm^3/g

Description: Coefficent that relates emission rate to exposure.

Definition: 10^(-9)

C is OK

S1 is undefined.
Do you want to define it? [Yes]:

Name of Variable? [S1]:

Title: First source term

Units: g/sec

Description: Strength of first source.
```

```
Definition: 10

S1 is OK
?        : Undefined.
Do you want to define it? [Yes]:

Name of Variable? [S2]:

Title: Second source term

Units: g/sec

Description: Strength of second source.

Definition: 0.7

S2 is OK
?         : Undefined variable
Spy is undefined.
Do you want to define it? [Yes]:

Name of variable? [Spy]:

Title: No.of sec.in a yr.

Units: sec/yr

Description: Coefficent relating sec. to yr.

Definition: 60*60*24*365.25

Spy is OK
?        : Undefined variable
D is undefined.
Do you want to define it? [Yes]:

Name of Variable? [D]:

Title: Slope of damage function

Units: D/per-ug-m^3/yr

Description: Slope of damage function for XYZ.
```

```
Definition: 0.0025

D is OK
?        : Undefined variable
P is undefined.
Do you want to define it? [Yes]:

Name of Variable? [P]:

Title: Population

Units: People

Description: Number of persons exposed to XYZ.

Definition: 1500

P is OK
Exd is OK
>

what EXD

Exd            : HEALTH IMPACT            (XD/YR    ) = 1.2662
```

If at this stage I list the model I have built by saying "WHY EXD", I get the following "self documented" model description:

```
Why EXD
Exd            : HEALTH IMPACT        (XD/YR    ) = 1.2662
Description: HEALTH IMPACT OF XYZ IN EXCESS DEATHS PER YEAR.
Exd = C*(S1+S2)*SPY*D*P


C              : EXPOSURE COEFFICENT (UGM^3/G  ) =      1p
S1             : FIRST SOURCE TERM    (G/SEC    ) =      10
S2             : SECOND SOURCE TERM   (G/SEC    ) = 0.7000
Spy            : NO.OF SEC.IN A YR.   (SEC/YR   ) = 31.56M
D              : SLOPE OF DAMAGE FUNCT(D/PER-UG-
                 M^3/YR)                          =  2.50m
P              : POPULATION           (PEOPLE   ) =   1500
```

Now suppose that after extensive interactions with appropriate experts we are able to obtain subjective estimates of the uncertainty attached to these coefficients. Subjective distributions for C, D and P have been elicited and are shown in Figure 2. S1 and S2 are estimated to be log-normally distributed, with geometric means of 10 and 0.7 and geometric standard deviations of 2 and 10, respectively. We can easily modify our existing DEMOS model and perform a fully probabilistic analysis as follows:

```
S1:=lognormal(10,2)

s1 is OK
>S2:=lognormal(0.7,10)

S2 is OK
>D:=fractiles[0,.00198,.0022,.00235,.00245,.0025
,.00256,.00261,.00266,.00285,.0035]

D is OK
>P:=fractiles[1300,1440,1470,1485,1498,1500,1510
,1518,1523,1540,1700]

P is OK
>C:=fractiles[2e-11,1e-10,3e-10,6e-10,8e-10,1e-9
,1.2e-9,1.4e-9,1.8e-9,2e-9,1e-8]

C is OK
>cdf EXD
```

Finally, we can explore alternative model formulations by simply changing the algebraic expressions and looking at the results.

The version of this paper presented verbally at Kansas City dwelt at some length on an example drawn from an actual risk analysis performed by the author and his colleagues on a problem involving long range transport and possible human health effects from sulfur air pollution from coal-fired power plants. Interested readers can find details on this work elsewhere (12).

Risk Assessment Tools as a Framework for Discussion Among Contesting Parties

Many of our institutions for risk management are explicitly adversarial. Different parties with different interests argue things out in institutional environments that are heavily based on legal models. Such procedures work in the sense that they usually produce "answers". Some would argue that they do rather less well in adequately identifying and dealing with scientific reality. Many would argue that they are inefficient and give rise to unnecessary controversies and tension.

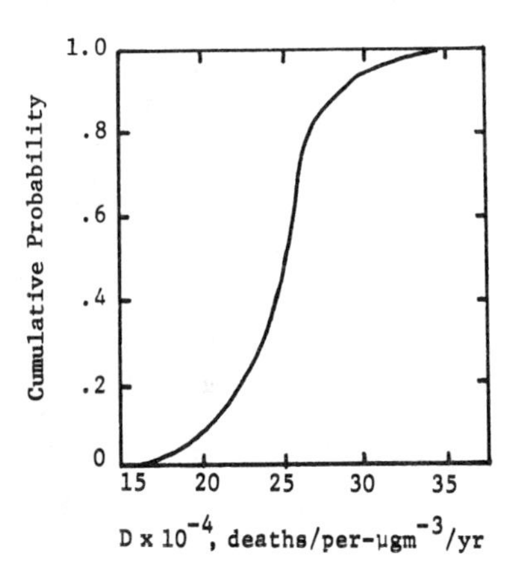

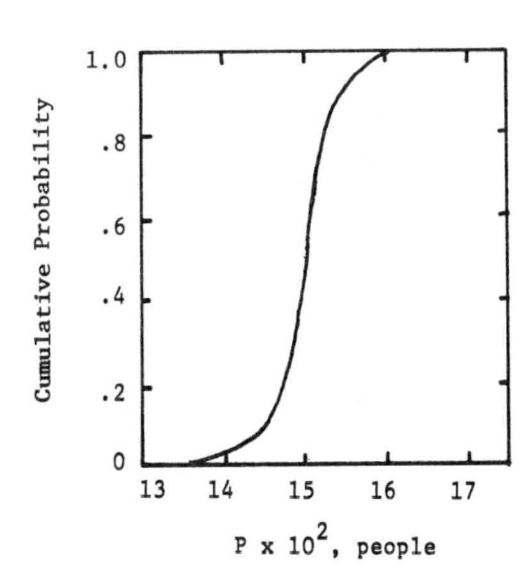

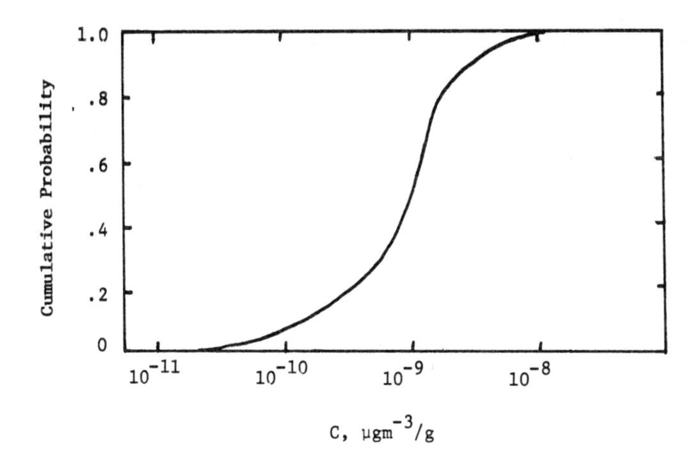

Figure 2. Hypothetical subjective distributions elicited from experts for use in the sample problem discussed in the text.

Right or wrong such concerns have increasingly lead people to suggest that we need to begin to experiment with our institutions for risk management, seeking, among other things, less adversarial more consensus-based approaches to risk management.

What is the possibility that "good" quantitative risk assessment ...that is risk assessment that strives to meet the criteria we outlined above in Section 2... might provide a framework for discussion among contesting parties in such a less adversarial more consensus-based approach could develop? On the positive side such an analytical framework might help the parties to quickly focus on their areas of agreement and disagreement. Since it would provide analytical results that treat issues of value parametrically the different parties could all use the same models and results to draw their various separate conclusion. Issues of value and issues of fact would be more clearly differentiated. Because uncertainty would be explicitly characterized and dealt with, there might be less tendency to use analysis as an adversarial dub since all the parties would find it easier to understand the implications of alternative model formulations and assumptions. In short, by providing participants with a "good" analytical framework that could serve a basis for deliberation, risk management decision-making might be made more "rational".

But, would it ...and do we want that? I don't know if it would. I suspect in many cases it could help considerably and for this reason I'm a strong advocate of running some well conceived experiments. But, at the same time, I have my doubts. Sometimes parties talk past each other, confuse issues of value and fact, read different meanings into the same results, and intentionally misunderstand the science because that appears to be the only way to reach a compromise solution. If a common analytical framework forced everyone to directly face and debate the real issues ...such as how much should society and individuals invest to save various kinds of lives in various circumstances... we might find ourselves deadlocked more often. Then again, focusing more often on the real issues might clear the air. We won't know until we try.

Acknowledgments

This work was supported in part by grant #IST-8112439 from the Division of Information Science and Technology of the U.S. National Science Foundation and in part by the Health and Environmental Risk Assessment Program of the U.S. Department of Energy. I thank T. Mullin and M. Henrion for their advice and P. Cibulka, I. Nair and E. Morgan for their assistance.

Literature Cited

1. Tversky, A.; Kahneman, D. Science 1974, 185, 1124-31.
2. Slovic, P.; Fischhoff, B.; Lichtenstein, S. Science 1974, 185, 181-216.

3. Morgan, M.G. Science 1978, 201, 971.
4. Keeney, R.L. Operations Research 1982, 30, 803-981.
5. Morgan, M.G. "The Role of Decision Analysis and Other Quantitative Tools in Environmental Policy Analysis: A Tutorial", OECD Environment Directorate, OECD Paris, 1983.
6. Henrion, M. Ph.D. Thesis, Carnegie-Mellon University, Pennsylvania 1982.
7. Morgan, M.G. IEEE Spectrum 1981 November, 18, 58-64.
8. Henrion, M.; Morgan, M.G. "A Computer Aid for Risk and Other Policy Analysis", preprint journal article, 1983, Department of Engineering and Public Policy, Carnegie-Mellon University, Pennsylvania.
9. Henrion, M. "DEMOS User Manual", Department of Engineering and Public Policy, Carnegie-Mellon University, Pennsylvania, 1982.
10. Henrion, M. in "Design Policy"; Design Council Books, London (in press).
11. Henrion, M. Proc. of IEEE International Conference on Cybernetics and Society 1979.
12. Morgan, M.G.; Morris, S.C.; Henrion, M.; Amaral, D.A.L.; Rish, W.R. "Technical Uncertainty in Quantitative Policy Analysis - The Sulfur Air Pollution Example", preprint journal article, 1983, Department of Engineering and Public Policy, Carnegie-Mellon University, Pennsylvania.

RECEIVED October 14, 1983

9

Use of Risk Assessment and Safety Evaluation

VIRGIL O. WODICKA

Consultant, 1307 Norman Place, Fullerton, CA 92631

The Scientific Committee of the Food Safety Council has developed and recommends a systematic procedure for evaluating the safety of food components. After careful selection and characterization of the substance to be tested and estimation of the human exposure pattern, testing starts with acute toxicity. Next come parallel tests for mutagenicity and studies of metabolism and pharmacodynamics. If the material is not mutagenic and metabolites are known to be safe, testing can stop; otherwise come tests for subchronic toxicity, including teratogenesis and reproductive effects. Chronic toxicity testing follows if necessary (according to criteria given). Decision after subchronic or chronic toxicity is based on extrapolation to low dose of the dose/response curve. Translation of the safe dose to man is based on mg of test substance/Kg body weight. Judgement can then be applied by a safety factor, considering all the evidence.

The problems of safety evaluation have been receiving increasing public attention and concern in the last decade. The concern has tended to focus on chemical safety because exposure to a wide variety of pure chemicals is new, and we tend to worry more about new things.

Exposure to chemicals comes mostly through four channels: air, water, food, and occupational exposure. In that the approach presented here originated in the Food Safety Council, its focus is on food, but it should be clear that the principles are generalizable, and are specific to food here only to avoid the cumbersomeness of language involved in being general.

Formal testing to assess safety has been the province of the toxicologist. Training for this work has been mostly in the province of the medical schools. In medicine, the discovery of

0097–6156/84/0239–0131$06.00/0

sulfa-drugs, antibiotics and vaccines has shifted the focus from communicable diseases to the degenerative diseases of old age. In similar fashion, emphasis in toxicology has moved from acute poisoning to the consequences of lifetime exposure to low levels of various substances. Even now, however, most teaching in the schools stresses the diagnosis and treatment of acute poisoning, and there is not a procedure generally agreed upon for evaluating chronic exposure.

The Food Safety Council

The relatively turbulent discussions of food safety in the last generation led some of the leaders in the food industry to form the Food Safety Council, with a Board of Trustees drawn from both public and private sources, to generate and propose a system of safety assessment that might serve as a nucleus around which to crystallize a system of general applicability. It formed a Scientific Committee to generate the scientific component of this system, and the work of this committee is the basis of this chapter.

The Committee developed and published a testing sequence organized into a decision tree in which each step integrated and applied the information previously gathered in reaching a decision to accept, reject, or further test the material in question. (1) Comments on this proposal were actively solicited, and two of the chapters of the proposal were then revised and again published. (2) As might be guessed, one chapter revised was the one on genetic toxicology, which was the newest field covered and still in active development and change. The other was risk assessment, about which more later.

Figure 1 shows the decision tree recommended. Details of the testing steps will not be given here because they are not relevant to the current issues. Suffice it to say that at various points in the sequence, a decision must be made to accept, reject, or test further. The decision hangs on the establishment of the risk level offered by the test substance and the degree of social acceptability of that risk. The report does not deal with the establishment of a socially acceptable risk level because this is more than a scientific issue; the report deals only with the estimation of the risk offered. The two points in the testing sequence of primary relevance here come after subchronic testing and after chronic testing.

Traditional Methods--Safety Factors

Toxicology is based on the assumption (amply supported by experience) that essentially every substance is harmful in some way at some level of exposure. The task of the toxicologist, then, is to determine the form of the damage induced and the level of exposure at which it appears. This all seems very logical, but the point that does not register with most people is that <u>every test that satisfies a toxicologist shows harm to some of the test animals.</u>

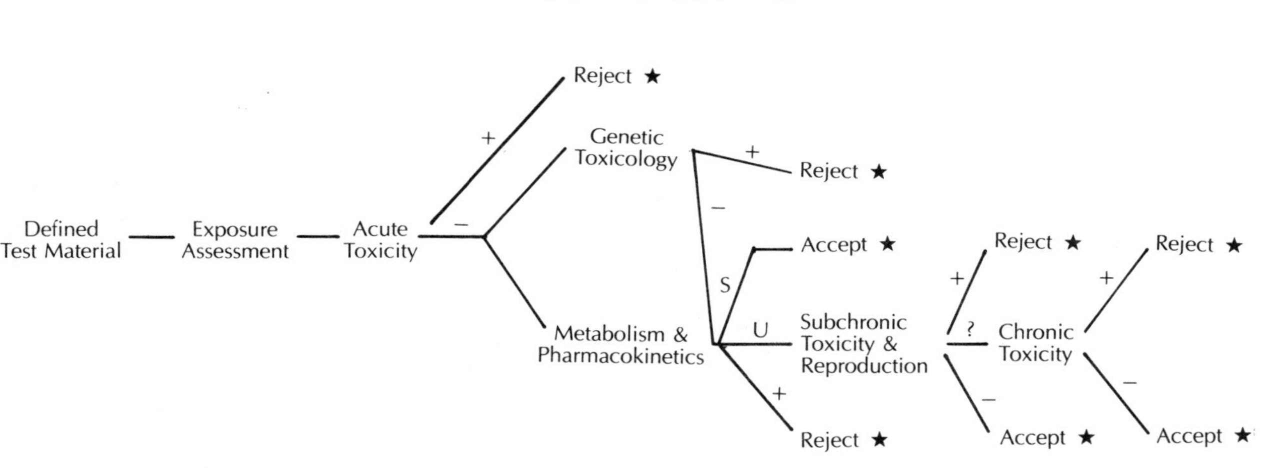

+ = presents socially unacceptable risk
− = does not present a socially unacceptable risk
S = metabolites known and safe
U = metabolites unknown or of doubtful safety
? = decision requires more evidence
★ = see Summary, page 5.

Figure 1. Decision Tree for Safety Evaluations.

Most of the essential nutrients will harm the animals if fed at a high enough level, and many of them will be fatal. Some are also carcinogenic. The point here is the paramount importance of dose. This point is well recognized by toxicologists but not by many others, including some other scientists, even life scientists and medical specialists.

The importance of dose level was well understood by the scientists working in food safety up to, say, twenty years ago. It was recognized in the procedure they evolved for safety evaluation. In either subchronic or chronic toxicity assessment, they exposed animals to several dose levels and established from their observations the highest dose level that resulted in no harm to the animals. Conscious of the many uncertainties, statistical and other, that underlay this point, they did not accept it as the safe dose level but adjusted it by a safety factor. This is standard engineering practice, including the terminology. The engineer designs a process or a structure, taking into account all known factors affecting performance, and then adds in a bit more protection to provide for the factors he did not and probably could not know about. This is called the safety factor but might be more accurately termed the factor of ignorance.

This is the procedure that has been and is still used by the Joint Experts Committee on Food Additives (FAO/WHO) in evaluating food additives and contaminants. It is also used by the Food and Drug Administration for food additives, color additives, contaminants, natural toxicants, or any other identifiable substance in food. It is built into the Code of Federal Regulations for such use. (3)

On a world-wide basis, the safety factor has been set on an _ad hoc_ basis as a matter of expert judgement on the basis of all the available relevant evidence. In the United States, it has usually been set at 100, and this is also possibly the most common factor world-wide.

There was originally a scientific rationale for the factor of 100, but the evidence supporting it was meager, much of the corresponding evidence since does not support the assumptions, and it leaves out some important sources of variation in effects. Accordingly, in today's world, it is better to look at it as just the crude factor of 100 with only the support of experience.

The experience has been good. The only instance I know of in which a food additive has produced damage is the use of cobalt salts in beer as foam stabilizers. Illness and some deaths resulted in people who consumed up to five gallons of beer a day, so the use was banned.

The overwhelming preponderance of direct food additives now permitted by law was built into regulations in the early 1960's after the amendment of the Food, Drug and Cosmetic Act in 1958 that provided for such regulations. These regulations were issued to cover substances already in wide use in foods at that time which were deemed to be not "generally recognized as safe." (With trivial exceptions, every component of food is legally a food additive unless it is generally recognized as safe for its intended use, by experts qualified by scientific training and experience to evaluate its safety.) (4) This means that with few exceptions, the direct food additives now in use have been in the food supply for at least 20 years. Most of the substances generally recognized as safe have been in the food supply for as many centuries. We are now beginning to develop questions about such components of food as salt and fats, but these are far from resolved and not relevant to the present discussion.

To add a little more perspective to the factor of 100, it should be recognized that few of the essential nutrients would qualify on this basis. If they were consumed at only 1/100th of their toxic level, we would have severe deficiency of at least several of the vitamins and many of the minerals.

All in all, it would seem that the safety factor approach has served us well and is amply conservative. Why should anybody consider changing it?

Cancer, a special case

One of the rocks on which this approach breaks is that of cancer. No responsible toxicologist would suggest settling for a safety factor of 100 on carcinogens. On the other hand, we now know that we are eating a variety of carcinogens every day. While cancer is a leading cause of death, most of us do not die of cancer, but all of us eat carcinogens. Just a few examples will make the point: There are dozens, perhaps hundreds, of references on the carcinogenicity of oxidized fat. Vitamin D_2 has been reported carcinogenic. (5) There are recent indications that at least some of the condensation products formed in the cooking of meats are carcinogenic. Something other than exposure must be involved. How do we draw the line between substances that may be tolerated though carcinogenic at high doses and those that must be avoided as much as possible? The safety factor approach could be used but is completely dependent on the judgement of those who choose the factor. There is no systematic approach to decision-making in this area.

Problems with safety factors

A more philosophical issue arises in this same context. The safety-factor approach assumes that at or below the no-observed-effect-level, there is a threshold below which nothing unpleasant happens. There is a strong body of opinion that carcinogens cannot have a threshold. A more objective view of biology would suggest that it is just as rash to assume that non-carcinogens always have a threshold as to assume that carcinogens never do. This is an unnaturally simplistic dichotomy. The discouraging aspect of this problem is that the exposure levels at which effects are found, be they cancer or anything else, are usually so far above real life exposure levels that there is no way now known or even suggested for discovering what happens in real life. Extrapolation downward of dose/response curves using any reasonable model gives incidence rates at real life exposure levels that are far below the sensitivity of any present or expected technique in epidemiology or requires animal tests using impossibly large numbers of animals to validate the estimates. In other words, there is no way now in sight to settle this question experimentally.

The next question we must look at is the nature of food additives. The cyclamate history demonstrates that by issuance of a regulation in the Federal Register, the Commissioner of Food and Drugs can convert a substance from the status of "generally recognized as safe" (GRAS) to a food additive. This same thing could happen with any closely characterizable substance in the food supply. Most people would agree that if there is a substantial doubt of safety, this should happen. Natural origin is an irrelevant consideration. After all, the Borgias did a fine job of estate-building before there was a synthetic chemical industry. As indicated earlier, few direct additives have been authorized since about 1962. Some of the recent ones have been the cell walls of yeast, poly-dextrose, and the organisms that produce the enzymes which convert glucose to fructose. The prospect is that most of the additives that are proposed in the future will not be exotic xenobiotics but concentrates or isolates from natural materials, microorganisms or their products, plant varieties resulting from DNA modification, or other variants of substances long in use. Many of these would be useless at levels 1/100th of their no-observed-effect level. For example, there has been talk of using L-glucose as a sweetener that is not metabolized and therefore contributes no calories. Such a material would have little value at levels less than 10% of the diet. What safety factor would one use? How choose it? How justify it? How would one handle a cereal grain from a variety that fixes nitrogen? In these days when retinoids are fashionable, how does one handle a vegetable treated with an enzyme that greatly increases its level of carotenoids?

There is also a problem from the statistical viewpoint. The no-observed-effect-level has a confidence interval that varies inversely with the number of animals in the test group. In a group of 50 animals, a finding of zero incidence may be statistically indistinguishable from a finding of one or even two, depending on criteria. Using the safety factor approach, the difference in permitted level between that resulting from a zero incidence and that resulting from an incidence of one can be large, indeed, even though they are statistically indistinguishable.

Another facet of the statistical problem is that this approach takes no account of the slope of the dose/response curve or of its curvature. It really uses only one point on the curve, no matter how many have been observed. The tacit assumption is that the safety factor is so large that it swamps differences in potency as reflected in slope. This obviously implies that the true safety factor is correspondingly variable.

A better way

This then leads to the question, is there a better way? The Scientific Committee of the Food Safety Council addressed this question and decided that a better approach is risk assessment, carried out by extrapolating the dose/response curve to low doses.

The next question to be addressed was that of the mathematical model to be used for the extrapolation. Most particularly, would one model do for all effects or was more than one required? This is obviously particularly a problem with cancer. Various models have been proposed for cancer, but there has been little consideration of the use of dose/response extrapolation for effects other than cancer; the safety factor approach is assumed adequate. For reasons given above, the Committee did not agree.

Review of a number of curves (40-50) on a variety of chemical classes and end effects gave no obvious reason for choosing a model based on the end effect. Most of the dose/response curves for all effects showed curvature, some more than others, but it became obvious that a model providing for curvature was the best choice. In that a straight line is a special case provided by the choice of the curvature parameter, this is accommodated by a model that handles sharply curved responses in other cases. The Committee looked at the probit model, which has been used widely in nutrition and pharmacology. The Committee felt, however, that this model could underestimate risk. The one-hit model had strong support from some workers for cancer. Carlborg, however, studied all the carcinogens on which he could find enough data to analyze and found that the one-hit model was not a good fit for any of them. (6)

The Committee then looked at the Poisson model but found problems with its discrete nature. It then modified this to make it continuous with a gamma model, for which it opted in its first publication. In the time between the two publications, the Committee studied the question further and found that most of the time, the one-hit model gave the highest probability of effect at a given low dose, the probit model gave the lowest probability, and in between fell the Weibull, the Armitage-Doll, and the gamma. The Weibull has a certain amount of intuitive appeal because it is the chief model used in reliability work in engineering--the prediction of time to failure of products. It is adaptable to the inclusion of time in the equations but does not require it. In the end, the Committee did not restrict its recommendations to a single model but suggested exploration of several, with the choice resting largely on goodness of fit but subject to influence from biological considerations such as the quality of the evidence. Obviously on screening tests, such as the many sponsored by the National Cancer Institute, where there is only one dose or two at most, no model choice is possible because there is no way to estimate curvature. The one-hit model is the only option, and this almost inevitably results in a ban.

Figures 2, 3, 4, and 5 show how the models differ in their treatment of the data. In each case, the plot in a shows the observed range, and the plot in b shows the extrapolation to low dose. The curves are determined analytically by standard techniques, not fitted by eye. These represent a sampling from 14 substances for which parameters are tabulated in the report. The pattern is much the same for all, regardless of the effect or the chemical nature of the test substance.

Problems with Extrapolation

Critics of dose/response extrapolation point out that the choice of model makes relatively large differences in the estimated probability at a given low dose. This is true, especially if the extreme models are included in the comparison. If only the Armitage-Doll, the Weibull, and the gamma are included, the differences are not too impressive.

It is true that they can be as large as two orders of magnitude. The spread is dependent, however, on the length of the extrapolation. For cancer, where attention has focused on the dose giving a probability of effect of 10^{-6}, there is often a considerable difference. For other, less scary effects, a higher probability can be tolerated, and the spread is less. All the middle models usually give an acceptable fit within the observed range, so a choice cannot often be made on this basis. In this connection, it should be remembered that a risk of cancer of 10^{-6} is equivalent to an increased death rate of about 3 per year in the total U.S. population. This, however, is a conditional probability. This risk occurs only if everybody in the U.S. consumes the calculated dose every day for a lifetime. This must then be multiplied by the exposure probability to give the real life estimate. In other words, an uncertainty as large as two orders of magnitude gets lost in the cushions.

Application to man

The next point to consider is that the extrapolation of the dose/response curve, strictly speaking, gives only the probability of the effect in question in the test animal species. This, of course, is not the desired answer. There still remains the conversion of the finding to man.

The traditional method, used widely in nutrition, pharmacology, and toxicology, is to assume the equivalence between species of doses expressed as milligrams of test substance per kilogram of body weight. This method was adhered to by the Scientific Committee of the Food Safety Council and is implicit also in the traditional safety factor approach.

There are critics of this approach, however. One alternative method is to assume equivalence of dose expressed as milligrams of test substance per square centimeter of body surface. (The body

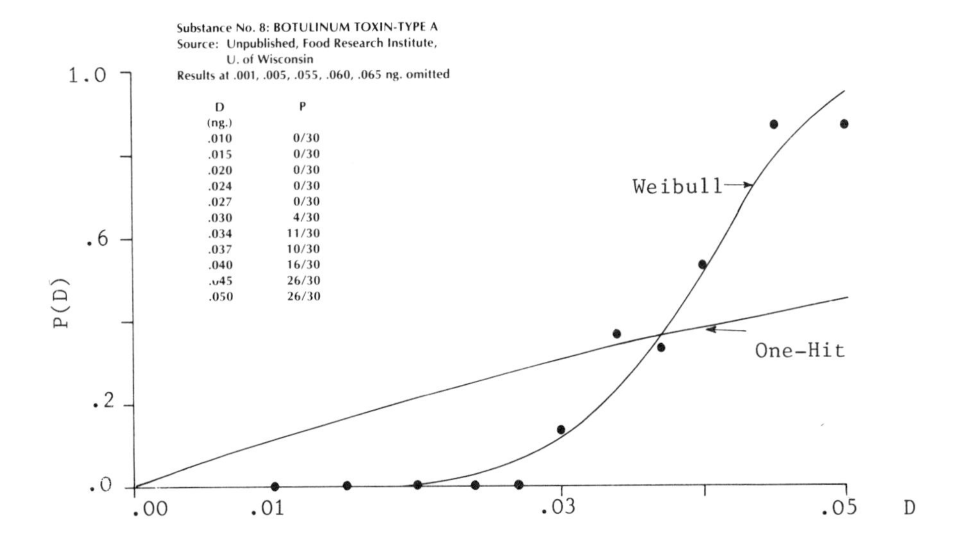

D (ng.)	P
.010	0/30
.015	0/30
.020	0/30
.024	0/30
.027	0/30
.030	4/30
.034	11/30
.037	10/30
.040	16/30
.045	26/30
.050	26/30

Figure 2a. Dose/response curves of best fit in the observed range for bladder tumors from sodium saccharin. (Gamma and Armitage-Doll models too close to Weibull to distinguish.)

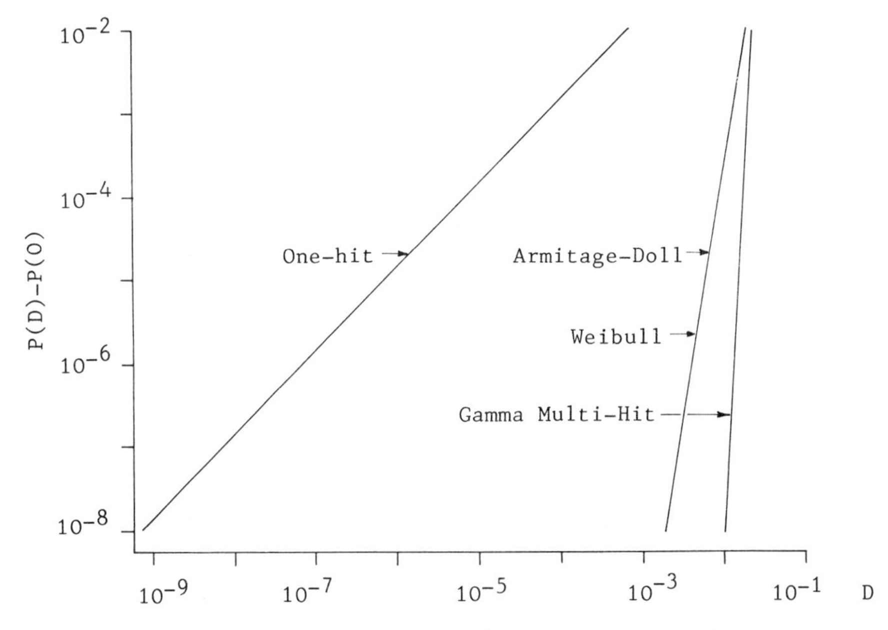

Figure 2b. Projection of dose/response curves for sodium saccharin to region of low dose.

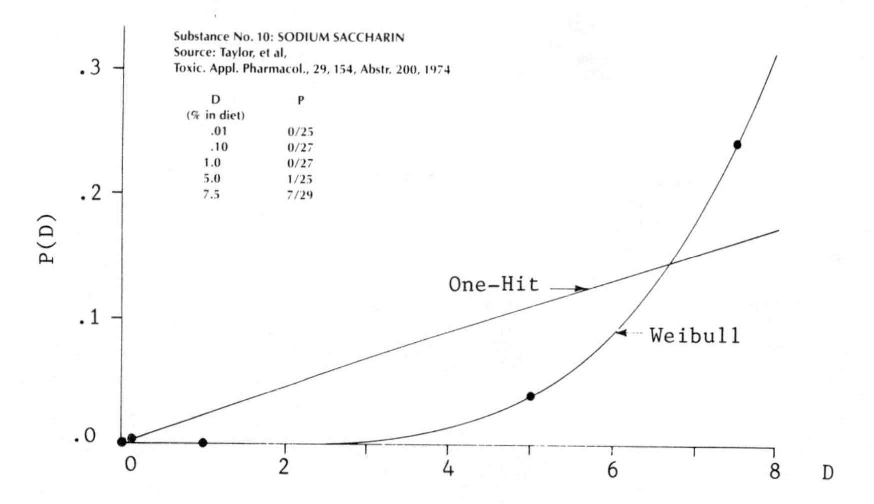

Figure 3a. Dose/response curves of best fit in the observed range for rat thyroid carcinoma from ethylene thiourea. (Gamma and Armitage-Doll models too close to Weibull to distinguish.)

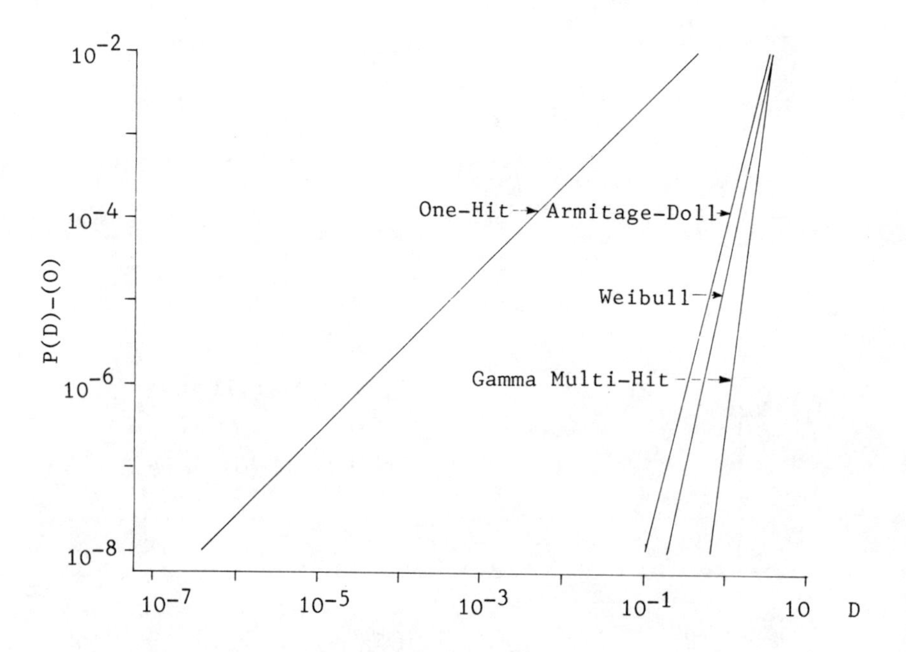

Figure 3b. Projection of dose/response curves for ethylene thiourea to region of low dose.

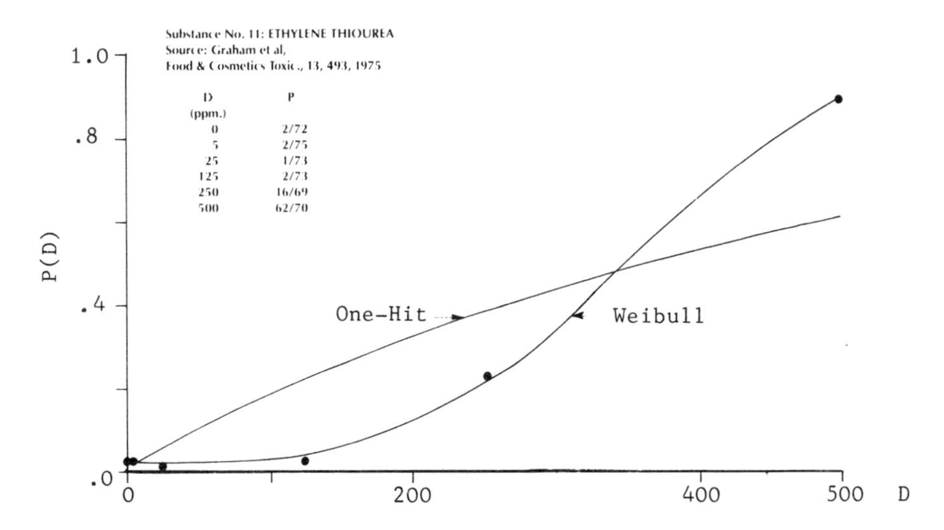

Figure 4a. Dose/response curves of best fit in the observed range for mouse hepatoma from DDT. (Gamma and Armitage-Doll models too close to Weibull to distinguish.)

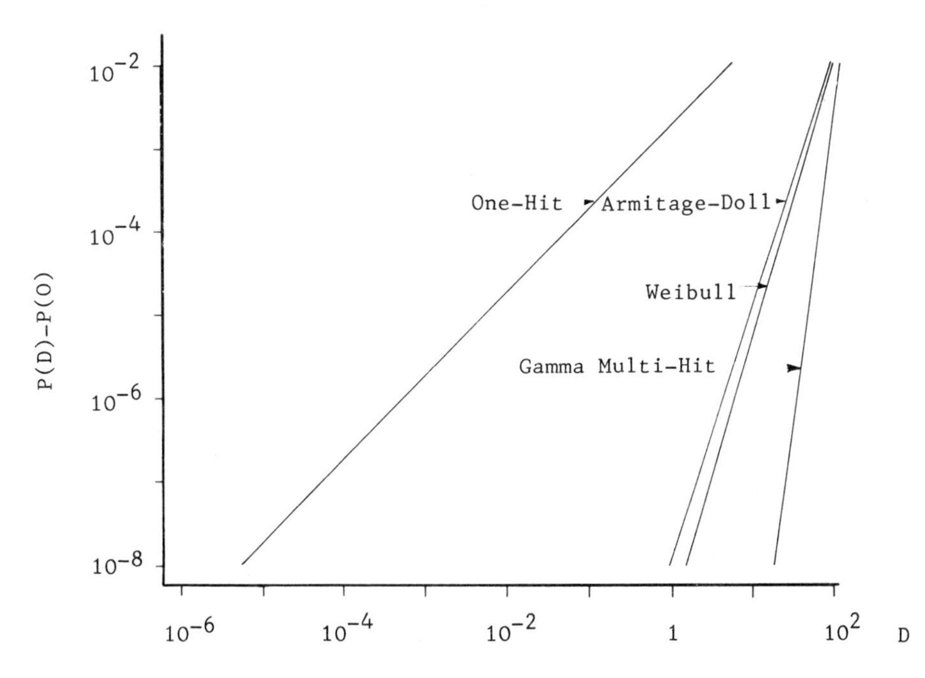

Figure 4b. Projection of dose/response curves for DDT to region of low dose.

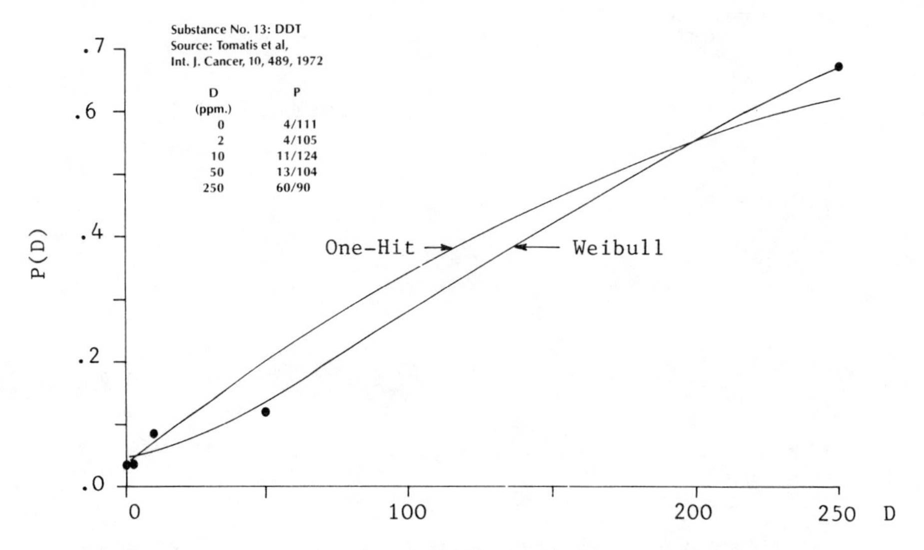

Figure 5a. Dose/response curves of best fit in the observed range for mouse deaths from Clostridium botulinum toxin. (Gamma and Armitage-Doll models too close to Weibull to distinguish.)

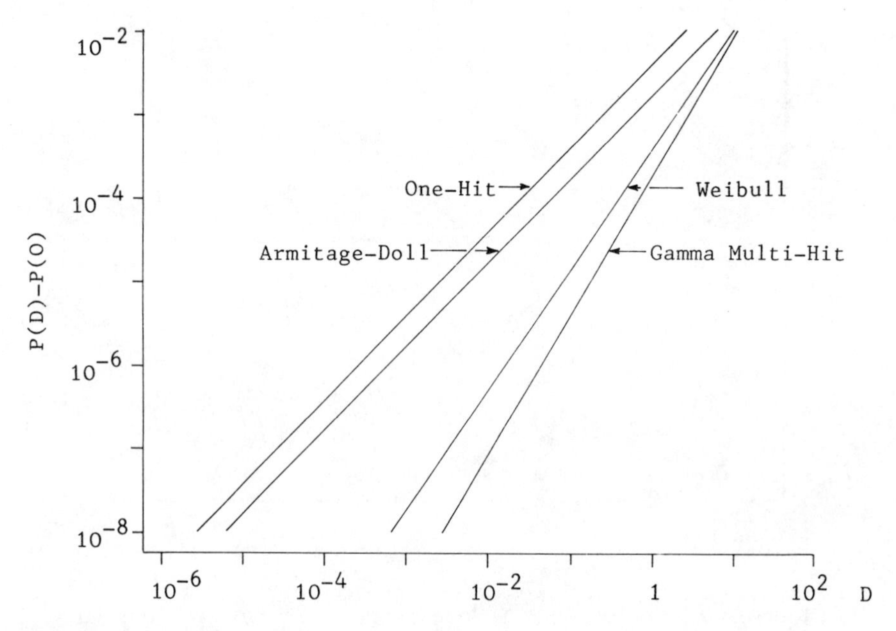

Figure 5b. Projection of dose/response curves for Clostridium botulinum toxin to region of low dose.

surface is usually calculated from weight, so no new measurements are needed; they are not related by a simple factor, however.) If this method is used, there is usually no change in the test design.

Still another method is to assume equivalence of concentration in the diet. In other words, parts per million for a man is the same as parts per million for a mouse in its effect. This method has profound implications. When conversion is on the basis of mg/kg, the concentration in the diet is altered every week or at least every two weeks to adjust for the change in weight of the animal and changes in food consumption. When equivalence of concentration is assumed, the concentration of test substance in the diet is held constant throughout the test.

It would seem logical to assume that the value that really counts is the number of molecules of test substance per target cell. This value cannot be used directly because before the test is run, it is not known what the target cells will be; that is one reason for running the test. This value is at least roughly approximated by the mg/Kg approach. Perhaps its weakest assumption is that the target organ, whatever it is, grows at the same rate as the animal as a whole.

When the concentration of test substance is held constant, however, the number of molecules per target cell is obviously highest at the start of the test. Food consumption per unit animal weight is highest then, and vital organs are smallest. There is the added factor that young animals do not have their enzymes and other defense mechanisms developed at the start of the test. This is true if the test starts at weaning, which is common, and it is even more true if the test starts at conception, which many people are now advocating. The dose in this case cannot be converted to mg/Kg by simple calculation because it is highest at the beginning and then decreases, at first rapidly and then more slowly, until the end of the test. The number of molecules per target cell may drop by a factor of ten or more.

Each of these methods of conversion has its strong supporters, and each is supported by a scientific rationale that sounds convincing if it is considered alone. There are even other methods, such as total weight consumed in the entire test and total weight consumed per kilogram of body weight. At present, there is really no strong evidence to support a firm choice. The variation in dose for man as calculated by the various methods is as great as that between mathematical models for dose/response extrapolation.

Summary

Given all the uncertainties, it may seem that the very laborious and expensive tests for assessing safety may be no better than throwing darts at a board full of numbers. It really is not quite that bad, and in any event, it is the only game in town. It is very much to be hoped that improved understanding of fundamentals will make it possible to do much of this work in vitro, which would be much faster and cheaper. Unfortunately, that possibility is not yet in sight.

The problem is that we are measuring effects with a meter stick, not a micrometer. When the media seize on one of these calculations and shout that 7,392 Americans will die of cancer or kidney necrosis or brain damage or whatever from exposure to substance X, this is a misuse of the calculation, no matter how it is done. We do not have the capability of estimating risk with that degree of precision, and we do not have the capability in epidemiology of knowing how close we really came. The point is that we know that carcinogens, which have been much studied in the past generation, can vary in potency over a range of five or six orders of magnitude. Other forms of damage have not been studied as intensively, but we know that the range of potency of substances causing them is also wide, though perhaps not that wide. We must have a way of sorting the potent from the weak in deciding how to handle risk from exposure. The methods we have are capable of doing that if they are applied consistently and with knowledge of their capabilities and limitations.

Finally, it must be remembered that there is no method of feeding the data into a computer and coming out with an unequivocal answer. It must be recognized that the execution of tests is not perfect in even the best of laboratories. This is particularly a problem in chronic toxicity tests which run so long and use so many animals that the chances of mishaps are correspondingly increased. Also, pathologists differ in their interpretation of slides, depending on their training and experience. In other words, the analysis of any safety evaluation, using whatever systems are available, will inevitably require a large measure of expert judgement, and should really be done by more than one person if the outcome is critical. The system proposed by the Scientific Committee of the Food Safety Council gives guidance in planning and organizing the testing and particularly in decision-making, but the Committee has been well aware that no system or formula will do the whole job; expert judgement is required at many points along the decision tree, and most of all at the end.

Literature Cited

1. Scientific Committee, Food Safety Council; Proposed System for Food Safety Assessment; Food & Cosmetic Toxicol. 16, Supplement 2, 1-136(1978).
2. Scientific Committee, Food Safety Council; Genetic Toxicology; Food & Cosmetic Toxicology 18, 683-710; Quantitative Risk Assessment, ibid. 711-734(1980).
3. Code of Federal Regulations, Title 21, 170.22.
4. Federal Food, Drug, and Cosmetic Act, Section 201(s).

5. Titus, D.S.; Crickman, E.T.; Pace, M.J.; Kleinfelter, V.R.; &Gass, G.H. Estimated activity of oral vitamin D_2 at carcinogenic dose levels. IRCS Med. Sci. 8, 286(1980).
6. Carlborg, F. W.; Dose-response functions in carcinogenesis and the Weibull model. Food & Cosmetic Toxicol.19, 261(1981)

RECEIVED October 25, 1983

10

The Need for Risk Assessment of Chemicals in Corporate Decision Making

FRED HOERGER

The Dow Chemical Company, Midland, MI 48640

Risk assessment is a complex and dynamic discipline, still in an evolutionary stage. The application of risk assessment within the chemical industry similarly has been evolutionary. Risk assessment is used in industry in several ways. Furthermore, industrial experience leads to several principles that are generally applicable in governmental and other sectors.

Chronology of the Growth of Risk Concerns

In the early period of chemical manufacture, roughly from 1900 to 1945, the industry focused on understanding and controlling the reactivity of chemicals. Properties such as flammability, spontaneous decomposition, autocatalysis, and acute hazards such as corrosiveness and other irreversible health effects from single exposures were characterized. Risk management practices were responsive to these types of data.

During this early period, the cornerstones of industrial toxicology and environmental science were established. For example, Dow, du Pont, and Eastman Kodak established toxicology laboratories. And the use of biological oxidation for treatment of industrial wastes was pioneered in the 1930s.

In the major growth period of the chemical industry, roughly from 1945 to 1970, the application of risk assessment to chemical hazards also grew rapidly. In the field of toxicology

0097-6156/84/0239-0147$06.00/0

the first two-year dietary study in animals was completed in 1946. Extensive animal studies on drugs, pesticides and food additives became commonplace in the 1960s. Industrial hygiene, a discipline concerned with control of exposure potentials and other hazards in the workplace evolved as a new profession. Corporations instituted occupational health programs with medical staffs and facilities. And environmental science became a focus of attention with the emergence of studies on aquatic toxicity, bioconcentration, and persistence of chemicals.

During this growth period of the industry, risk assessment was largely formalized by corporate actions on a case-by-case basis and a professionalism involving data generation and scientific judgment. Relatively few professionals were involved, these coming from industry, academia, and government. As examples, a few of the groups contributing to these formalized risk assessments were the American Conference of Governmental Industrial Hygienists, for establishing exposure limits for chemicals in the workplace; The Food and Drug Administration which established residue tolerances for pesticides; and the American Society for Testing Materials which established analytical methodology.

Growth of the chemical industry has been more moderate since 1970. But during this third period, a combination of factors has increased almost exponentially the focus on risks associated with chemicals, health and the environment. These factors include advances in health and environmental technology, an increasing corporate responsibility toward health and environmental concerns, media spotlighting of new information and incidents, and increasing government and public involvement in risk decisions.

Advances in health and environmental technology during the 70s were truly dramatic. Analytical technology which formerly ferreted out chemicals in the ranges of tenths of a percent to parts per million now find chemicals in the parts per billion and parts per trillion range on a routine basis. Carcinogenic bioassays have increased during the 70s and are now augmented by a variety of short-term testing methodologies. Mutagenicity, teratology, reproductive effects and behavioral toxicology are only a few of the subdisciplines of toxicology and clinical medicine now receiving sophisticated attention. Epidemiology has become a byword for both industry and government researchers.

New information on the effects of chemicals has received widespread media attention. Bioconcentration of chemicals such as PCB and DDT in fish, new findings of carcinogenicity, and the occurrence of reproductive health effects, to name only a few, are topics which have been publicized in media of every type. Incidents ranging from transportation spills to the leaching of chemicals into ground water from waste disposal sites have similarly been publicized. The public has developed perceptions of risk based upon fragments of information rather than comprehensive risk determinations.

The government's role in risk determination has paralleled the concerns of the media and the public. Frequently, perceptions of risk rather than objective fact collection and risk assessment have been paramount in Congressional enactment of laws and in government regulatory decisions. Carcinogenicity has been the trigger for a whole host of government decisions, for example, controls on DDT, asbestos, benzene, saccharin; and even to enactment of the Clean Drinking Water Act.

Significantly, a strong trend toward government reliance on risk assessment occurred since 1976. The Toxic Substances Control Act of 1976 stipulates that regulatory control over chemicals shall involve decisions based upon a balancing of health and environmental risks with economic and social impacts. Congress, responding to a changing public mood, deferred FDA's proposals to ban saccharin. The Supreme Court, in overturning OSHA's standard for benzene, stipulated that OSHA must make a threshold finding of significant risk. FDA was directed to allow de minimis quantities of acrylonitrile in polymers rather than insist upon zero concentration.

Finally it should be mentioned that an increasing number of corporations have continued to expand their corporate policies toward the knowledge and perceptions of health and environmental concerns. For example, many company policies relate, in a comprehensive way, to product safety, workplace health and safety, and to disposal of chemicals.

Broad Principles of the Risk Assessment Process in the Corporation

Based upon the events cited in the preceding brief chronology, I believe there are two principles that influence the risk assessment process in corporations.

The first principle is that a combination of data and experienced judgment are necessary to assess risks in a rational way. The need for data on the intrinsic properties of a chemical is self-evident. The experienced judgment is perhaps less obvious. The seventies became the era of risk speculation. The public, government regulators, and plaintiff attorneys are familiar with the lengthy list of possible hazards from broad generic classes of chemicals. For a given chemical, the likely risks must be separated from the long speculative risk lists. The experienced judgment must integrate chemical and physical properties, use/exposure potential, toxicity and environmental fate information in order to target the risk assessment process toward real risks instead of trivial risks.

The second principle of the risk assessment process in the corporation is that it is a staged and recycling process.

Generation of new data leads to new conclusions and interpretations. An oncogenicity study may provide a negative indication of cancer potential and thus validate present exposure controls. On the other hand, a positive indication of carcinogenic potential may trigger both a reevaluation of exposure controls and the need for further data generation.

Another major source of the recycling of risk assessments comes from generic advances being made in our methodologies. As oncogenicity studies have progressed, we have seen a thrust from animal studies of 6 months exposure to studies involving lifetime exposure to qualitative short-term tests. And today, we see an emerging emphasis on mechanism studies to differentiate genetic from non-genetic carcinogenesis. As each methodology develops, it may be necessary to fill data gaps, and to update or revise the risk assessment of a given chemical.

The point is, a risk assessment is a snapshot in a time frame subject to an expanding data base and advances in scientific methodology, theory, and insight. Experienced judgment is a necessary component to target resources and to articulate priorities against the indiscriminate background of speculative risks.

Types of Corporate Decisions Based on or Relating to Risk Assessment

From an analytical viewpoint, corporate decisions reflecting health and environmental risk assessment are of two types:

1) those involving business decisions on projects, investments and ventures; and

2) those involving policy considerations such as long-term staffing plans, and employee, customer and community relations.

The degree of importance of risk assessment depends upon the specific circumstances. It depends upon the product mix of the company and the type of activity which is carried on, that is, whether the company is a manufacturer, a processor or a distributor.

Almost every company at some time is faced with the question of whether to continue or discontinue a product or activity based upon new findings or perceptions of risk. The more common problem is frequently faced in decisions on whether to adopt more stringent practices or standards in order to minimize risks. Frequently an investment decision is clouded by uncertainties, or perceptions of uncertainties in risk assessment, or by gaps in risk information. Commercial development of a new product or a new venture often hinges on a risk assessment and an evaluation of the potential for liability claims and litigation costs. The selection of a new plant site must be influenced by the determination of several kinds of risks.

Difficult decisions are often faced by management in determining the optimum degree of communication for new risk information. Usually new information on health effects comes in fragments so that perspective on risk is difficult to maintain, whether communicating to employees, to customers, to the government or to other public sectors.

Company Policy on Risk Determination. Either by careful design, by an evolutionary process or by default, a given company usually arrives at a number of significant policies and practices involving risk. Considering the events of the past decade and the continuing trends in risk concerns, an essential policy decision involves the assessment of staffing needs. A company must develop the technical and specialist resources:

- to maintain a current awareness of all information relating to the risks of its products and activities;
- to determine appropriate practices for minimizing risks;
- to disseminate appropriate customer information; and
- to work with customers and various government and public bodies in their assessment of practices and standards.

Almost as important is the decision on whether to develop resources for research and testing to obtain additional risk information. A company can respond to information generated by the various industry, academic and government sectors, or it can participate in the development of information. The latter course permits a more in-depth awareness of sophisticated technology at the professional interaction level.

Employees, customers, and members of local communities are aware of, and often highly sensitive to, information on the effects of chemicals. Policies tending toward candor, direct dialogue and mutual problem solving emerged during the 70s and will probably continue in the future.

The Interrelationship of Risk with Other Corporate Considerations. For purposes of this discussion, assume that the actual risks involved with a substance or a given activity are objectively judged to be quite low. Even so, a number of other factors must be taken into account.

It has become increasingly clear that the public perceptions of risk do not necessarily follow objective statistical assessments of risk. For example, several surveys have been conducted in which the public was asked to rank according to relative risk various industries or activities such as pesticides, nuclear power, airplane travel, smoking, skin diving and others. The public ranking did not correlate very well with statistical evidence and the expert judgment of engineers and scientists. It is interesting to note that use of asbestos insulation in hair dryers probably presents a trivial risk situation; however, it is unlikely that any company would today consider a reintroduction of such a practice.

Frequently the amount of effort required to demonstrate that a risk is trivial becomes very large. For example, one can hypothesize that any gasoline additive might have the same health effect characteristics as benzene. If this was being considered as a new chemical venture, the effort to put the low risk into perspective might be almost insurmountable. In 1971 NTA was voluntarily withdrawn from the market by major detergent manufacturers because concerns were raised about the safety of the material. Later studies resolved these concerns. In 1973, NTA-containing detergents were reintroduced into Canada and later approved for use in several European and South American countries. In the U.S., significantly more research and testing were done on NTA. Finally in 1980, after a 2-1/2 year review of NTA safety, EPA concluded that there was no reason to take regulatory action against the resumed production and use of NTA in laundry detergents.

It seems apparent that smaller volume products with less market potential than NTA or benzene and gasoline cannot justify the effort required when public perceptions of risk are greatly different from objectively determined risks.

The Distinction Between Risk Assessment and Acceptable Risk. Both corporate and government decisions on risk ultimately focus on levels of risk that are acceptable to society. Because of the widespread inconsistencies in risk levels achieved by the inclusion of political factors in the regulatory process and because of differences in laws, regulatory decisions show a pattern of inconsistency in the level of risk control actually achieved. These inconsistencies have increased the focus on methodology for estimating risks from various societal and industrial activities. It is important to emphasize that the desired goal of objective risk assessment should be that it is viewed as one of the tools for the decisionmaker and that no specific methodology leads to a single formula for a decision. Within the corporate setting, it is felt that management needs the best assessment of risk associated with the various uses of a chemical. The risk analysis can then be integrated with the other business and public policy considerations.

Data Generation and Risk Assessment Integrated into the Research, Development, and Commercialization Cycle of a Product

As research progresses toward a commercial venture, there are points in time when the importance of the chemical substance, formulation or the fabricated item must be evaluated. Although each company may have its own system for evaluation, more and more individuals become involved as the cycle progresses. Multi-disciplinary resources and talents are needed to deal with procurement, with the biological impact or potential hazards from the material, with the formulation or fabrication methodology that may be needed, and even with how it will be marketed and what segments of society will find it useful. Thus, industrial R&D involves managing a broad range of resources -- dollars, professional skills (people) and facilities (analytical equipment, pilot plants, etc.).

Industrial research and commercial development of a product are frequently viewed as being segmented into four stages:

Stage I -- Exploratory and Synthesis
Stage II -- Product and Use Characterization
Stage III -- Pilot Process and Field Development
Stage IV -- Commercialization

It is in Stage I -- the exploratory stage -- that scientists initiate effort on a problem, the solution to which will contribute an economic benefit to the company and fill a need that the user is able to identify and for which he is willing to pay. The scientist will be looking for new concepts, potentially useful compounds and new ways of modifying existing products, and he will be exercising the processes of innovation and invention.

Stage II represents a selection point. More resources are targeted on a given product and/or technology. Stage III is further targeting of resources -- more facilities, more disciplines, and involvement of those outside the company -- leading to commercialization in Stage IV.

Let's examine several different aspects of this four-stage process.

Progressively, the corporation proceeds from many ideas and chemicals to one chemical contained in one or a few formulations for one or a few uses. The knowledge of the potential use becomes more sophisticated through this progression. For example, from a concern for the simple property of tensile strength in a polymer the corporation will become more sophisticated and in turn examine other parameters such as brittleness, adhesive characteristics, light stability, and many more. The quest for a broad-use product frequently narrows down to a specific use.

Generation of toxicological data similarly goes through a progression. Newly synthesized compounds may be characterized only as to acute toxicity to the rat or be screened only for unique drug or pesticidal activity. But as the progression proceeds and commercial success becomes more likely, the toxicological data base and potential exposure are further considered and, where needed, more species are tested and longer range tests may be commenced. For example, early in Stage II, mammalian toxicological range-finding tests frequently are instituted. These studies are designed to determine the capacity of the material to cause injury from acute exposure if ingested, makes contact with the eyes or skin, or is inhaled. These types of studies are done to determine any significant degree of danger from incidental exposure to the compound. This information is needed for safe handling by chemists and chemical engineers, and is useful in the design of pilot plants. This information also will become part of the data package made available to the customer, to the government and to other segments of society.

Environmental data similarly goes through a sequence. Considerable insight on transport and fate characteristics can be gained from simple chemical and physical data collected in Stages I and II. Vapor pressure, water solubility, dissociation constants, hydrolysis and oxidation half-lives, along with a simple test for biological oxidation, permit informed judgments on environmental characteristics. Modeling studies, kinetic analysis of degradation processes and aquatic toxicity studies will be considered for those few large volume chemicals which have potential for large environmental release, are relatively persistent and exhibit relatively high toxicity.

Group Decisions on Product Review. In the latter part of Stage II, the first key product review involving many individuals takes place. Up to this point, decisions have generally been made by individuals personally involved in the product development. As a product becomes more identifiable, group review takes place and a management decision is made on whether to begin intensive development activity.

Suppose the compound shows little potential for environmental insult and has also proved to be very low in toxicity. In such a case, there is an opportunity to move more rapidly toward commercialization without further extensive testing, especially if the product is site-limited, is useful as an intermediate for the preparation of other types of materials, or if it has only minimal contact with man and the environment.

If, however, there is opportunity for high exposure to the compound or it has a high probability of entering into the environment, it may be necessary to do sub-chronic toxicity tests. Such studies may be of the ingestion type or of the inhalation type, or both, and may take three to six months. In addition, skin sensitization tests may be performed on guinea pigs. Concurrently, product evaluation research is intensified in the laboratories, and frequently potential customers are encouraged to evaluate the product in their selected uses.

As the chemical product moves toward commercialization, other major key review conferences are conducted. Committees of research, development and marketing people review the additional data and decide whether to obtain additional information, including health and environmental data, or go directly to the marketplace.

With a relatively few new chemicals, there may be potential for wide exposure to humans or the environment, and if the initial toxicity and environmental profile data indicates concern for certain effects, there may be need for an in-depth series of studies which could take three years or longer to complete. These studies would be chosen from a spectrum of animal tests such as studies of metabolic pathways, pharmacokinetic parameters, teratogenic, mutagenic and carcinogenic potential and reproductive competency. Normally the necessary tests are chosen by considering potential use and similarity to already characterized compounds. These decisions require professional judgment by highly experienced experts.

While the additional health and environmental testing is going on, much work in advanced product and process development is also being carried out.

The health and environmental data collected in the laboratory becomes the basis for defining operating controls and practices and for manufacture and use. Typically, the chemical manufacturer, and many users, initiate industrial hygiene monitoring and medical surveillance programs in the commercial phase.

It can be seen that data generation for risk assessment is an on-going process during development and commercialization of a new product. Once commercialized, a few chemicals grow to large volume (multi-million to multi-billion pounds per year) and frequently their utility results in uses not envisioned at product launch. Risk assessment continues with this growth.

Risk Assessment on Existing Chemicals -- Data Gaps, Exposure Assessment and Priorities

The inventory of existing chemicals, compiled in 1978 under the Toxic Substances Control Act, lists more than 55,000 chemicals in commerce. (Pesticides, drugs and food additives are excluded). Considering the rapid advances in health and environmental sciences during the past two decades, and the limitations on facilities, professionals skilled in testing, and monetary resources, it is obvious that not all of these existing chemicals can be tested and evaluated by a universal testing prescription.

As a generalization, most large chemical companies review the information base on their chemicals on a periodic basis in order to establish research and testing priorities. Absence of data, structure/activity similarity to chemicals with potent adverse effects, and potential for exposure are important considerations in setting priorities. Usually, a dossier of available information is prepared prior to establishing a testing program.

Similar practices have been utilized by the Chemical Industry Institute of Toxicology in establishing their testing programs and by the Federal Interagency Testing Committee which functions under TSCA.

It is becoming increasingly clear that estimates of exposure potential are important in developing testing priorities. Screening of groups of chemicals to select candidates for possible testing can rely on surrogates for exposure data:

- Volume of production
- Amount of environmental release
- Type of manufacture and use:
  - on-site
  - chemical intermediate or consumption use
  - Handled in close systems.
  - Dispersion uses.
- Environmental Fate Characteristics
  - Persistence
  - Bioconcentration
  - Media Transport

Blair and Bowman* have analyzed the volume profiles of the universe of chemicals. They found that 10% (3,796 substances) are produced in quantities over 1 million pounds, per year, but that 81% (31,699 substances) are produced in quantities less than 100,000 lbs/yr. They also reported that significant health and environmental data exist, or is currently being generated, on 21 of the 50 largest volume chemicals.

Currently, much of the testing to fill data gaps is targeted at the large volume or commodity chemicals. Considering that risk is a function of the toxicity (or other hazard) and the exposure, it is apparent that greater reliance on exposure assessment will be necessary as priorities are established for medium and small volume chemicals.

---

*Blair, E. H., and Bowman, C., Control of Existing Chemicals. Presented at the American Chemical Society National Meeting, Las Vegas, Nevada. March 31-April 1, 1982.

Sound Risk Assessment: Multidisciplinary and Staged

As can be seen from my comments so far, risk assessment in the corporation is frequently an informal process integrated into other aspects of projects, programs, and decisions. It is, however, becoming more formalized, along with needs for more widespread communication. It seems worthwhile to elaborate on two key aspects for sound risk assessment -- their multidisciplinary and staged nature. These key features have evolved from experience and seem equally applicable to the government and other sectors making assessments.

Risk Assessment -- A Multidisciplinary Process. Sound risk assessment must be considered as a multidisciplinary function relying upon data and experience from toxicology, epidemiology, clinical medicine and industrial hygiene or engineering representing exposure aspects of substances or agents.

The risk assessment should be carried out by a group of experts in these fields, making up a science panel. The health professionals who make up this panel may be employees of the company and may be called an industrial health board. In other cases, the panel may be employees of an outside consulting firm or may be academic consultants.

Operationally, the simplest version of multidisciplinary risk assessment would involve the preparation of a draft risk assessment by one or two scientists which would then be subjected to critical review by an academic consultant knowledgeable in the subject area.

Risk Assessment -- A Staged Process. Risk assessment is obviously complex. If the goal is to have a reasonably objective determination of risk, it is useful to view the determination as a sequential process and as itemized in Table I.

Table I. Risk Assessment Is a Staged Process

1. Perception of risk
2. Preliminary risk assessment
3. Proposed fact collection and research program
4. Peer review
5. Implementation of fact collection and research
6. Comprehensive risk assessment

Some event triggers a perception of risk:

> a new toxicological finding, a pattern of association events such as disease or wildlife injury, an analysis of data fragments, or a series of customer inquiries or complaints.

Such perceptions usually lead to a preliminary risk assessment which may be made on the basis of readily available facts. Frequently the available data are limited resulting in high degrees of uncertainty in the preliminary assessment. If it is deemed desirable to reduce such uncertainties, then a fact collection and/or research program must be planned. It is important to emphasize that any more refined risk assessment will depend upon the quality of judgment involved in the design of research and fact-finding projects. Interaction of several disciplines is important to maximize the utility of the results. Attention to the purpose and detail of protocols, anticipation of environmental concentration levels, and a host of pragmatic questions are essential. Typically, one or more experimentalists and a sponsor will draft the research program. Peer review by one or several experts can then upgrade the drafted program.

Fact collection and research results then set the stage for a comprehensive risk assessment. Some would argue that the staged process outlined here would delay management decisions. It may be that the fact collection stage would be longer; however, the greater clarity of the risk estimate would provide more precision in addressing alternatives and in many cases would avoid waste of resources, reduce the confusion and increase the degree of confidence and validity of the decision. Preliminary emphasis upon the staging needed to develop good science and fact would in reality work to shorten the time used in wheel-spinning and course changes and the providing of contingencies due to uncertainties.

Anticipation of Regulatory Decision

As federal and state involvement in setting health and environmental standards has increased during the seventies, corporations increasingly have had to anticipate the outcome of regulatory proceedings. For example, if new plants must be retrofitted to meet new standards, compliance costs escalate; or, the demise of a major use for a chemical may idle a plant and its workforce.

Predicting regulations is not an easy forecast to make, but the zones of uncertainty are decreasing. Some statutory guidance applies. For example, the Delaney Amendment requires a simplistic risk assessment, centering solely on the questions of whether a material is a food additive and whether it is carcinogenic in humans or in appropriate animal tests. A recent Supreme Court decision in the benzene case stipulates that OSHA must show a significant reduction in risk in order to justify a workplace standard. This infers reliance on risk assessment, but leaves open the question of significance. The Toxic Substances Control Act, in contrast, stipulates consideration of risks, but balances risks with costs and benefits.

During the 1977 to 1980 period, OSHA, the CPSC and to some extent FDA and EPA adopted generic regulatory policies for evaluating carcinogenic risks. These policies permitted us to predict the final outcome. But they had a major shortcoming -- they prescribed a recipe approach to risk assessment which precluded an evaluation of all the available data and an expert judgmental interpretation of the interrelating factors and their consistency with underlying scientific theory.

The author believes that risk assessment by a multidisciplinary group can relate a given risk situation to other risk situations that have been dealt with by regulatory agencies. However, the risk assessment is, at best, only one of numerous inputs to regulatory decisions.

Perspective for the Future

During the past few years considerable debate has focused on the uncertainties of risk assessments, the concept of making prudent assumptions, the concept of upper limit estimates, 95% confidence levels and other factors relating to the uncertainty of the data. Despite these debates, it has been necessary to make many decisions in both company risk management practices and the public regulatory arena.

It is important that risk assessment aim at determining the most probable estimate of risk. Utilization of a sequential process of multi-disciplinary risk assessment, with a focus on a comprehensive data base, will go a long way toward achieving this goal.

The decisions from top management, both in corporations and agencies, which commit to the development and use of sound risk determinations will become increasingly important.

As indicated earlier, our needs are sound data and technology for appropriately minimizing health and environmental risks. To accomplish this we need specialized skills, interdisciplinary interpretation and judgment.

Laws, regulations and the authority vested in government are now superimposed on our chemical development scheme. Legalistic and bureaucratic processes have a tendency to place burdens of proof foreign to the normal processes of scientific interpretation and the tests of consistency of the data as a whole.

Governments have articulated scenarios that involve new requirements for testing and criteria for decision-making. Many of these scenarios, when carefully examined, raise specters of testing ourselves to extinction.

We are today part of a great debate in risk management. This arena requires perspective on its complexities and the balancing of our important human, economic and natural resources.

RECEIVED November 4, 1983

11

Chemical Industry Perspectives on Regulatory Impact Analysis

RENÉ D. ZENTNER

University of Houston Law Center, Houston, TX 77004

This paper will address the current state of regulatory impact analysis in the health, safety, and environmental areas. In particular, it will discuss the present state of regulatory impact analysis and some evolving trends in the evaluation of environmental regulatory proposals. It will also set forth some principles being proposed by the chemical industry for application to regulatory impact analysis in the light of the current regulatory climate.

In preface, it is worthwhile to define "risk management" as it is understood to be used in this symposium. By "risk management" is meant the abatement of risk to a socially acceptable level, considering the elements of costs and benefits of the chances taken by individuals through exposure to potentially hazardous environments. Such abatement is achieved through determination and evaluation of regulatory alternatives and selection of the alternative that offers the greatest benefits at the least societal costs.(1)

By this definition, most risk management has occurred before a health, safety or environmental issue has come to regulatory attention. Responsible industrial enterprises determine and control the risks to which individuals are exposed through design and engineering of manufacturing plant processes, through product formulation, risk-reducing packaging and through warning labels. Thus, most risks are evaluated and managed by companies as an integral part of the management of the enterprise. It is only when a social determination is made that this level of management is inadequate that governmental intervention is proposed to further control individual exposure through imposition of regulations. Risk assessment is thus involved in both private and public decisions of how such exposure is controlled.

This paper will not deal with the scientific and technical aspects of risk assessment. Those aspects are extensively dealt with elsewhere, and are the subject of a growing literature of their own.(2) This paper will instead address evaluation of

0097-6156/84/0239-0161$06.00/0

regulatory devices intended to enable society to manage the hazards to which it is exposed as a result of such human activities as industrial production. In particular, it will present the views of the Chemical Manufacturers Association on the most recent Presidential Executive order mandating the use of cost-benefit analysis in regulatory rule-making.

The virtues and vices of various means for regulatory control of individual exposure to hazardous substances have been extensively addressed in the literature, and it is not the author's purpose to address them here. Instead, this paper will assume that American society has determined that such exposure will be regulated by various governmental agencies, authorized by appropriate legislation to do so. The task of those regulated, or those affected by the regulation, is to be sure that such regulation is the most effective means for achieving the object sought by the authorizing legislation. How that determination may be accomplished is the subject of this paper.

The goodness or badness of regulation has been debated by scholars, public officials and the regulated since serious regulation of American society began a century ago. The debate has traditionally been conducted between advocates of various regulatory devices, who supported the benefits of the regulation, and the subjects of the regulation, who complained about its costs. Because the benefits of the regulation generally have often been said to be felt by one class of society and the costs borne by another, it has seemed hard to reconcile these costs and benefits. Since February, 1981, however, American regulatory agencies have been asked to attempt to do so, using cost-benefit analysis. In the following discussion, this analysis will be discussed and its limitations examined.

Cost-Benefit Analysis as an Analytical Tool Cost-benefit analysis has been an established analytical tool for evaluating major public sector projects for almost a century and a half. Although such evaluations were no doubt conducted for efforts of this kind throughout human history, the modern literature of the method generally dates from 1844, with the publication of an essay, "On the Measurement of the Utility of Public Works" by Jules Dupuit, a French engineer. Dupuit introduced his subject by stating:

> "Legislators have prescribed the formalities necessary for certain works to be declared of public utility; political economy has not yet defined in any precise manner the conditions which these works must fulfill in order to be really useful; at least, the ideas which have been put about on this subject appear to us to be vague."

Those attempting to employ cost-benefit analysis for the evaluation of contemporary projects will recognize the same vagueness, incompleteness and inaccuracies experienced by Dupuit.(3)

Since that time, cost-benefit analysis has been employed for systematically developing useful information about the desirable and undesirable effects of public sector programs or projects. It has been described by some writers as the public sector analog to the private sector's profitability analysis: the former attempts to determine whether social benefits of a proposed public sector activity outweigh the social costs whereas the latter attempts to determine whether the private benefits, e.g. revenue, outweigh the private costs. The method has been applied extensively to such diverse areas as studies on air pollution control, consumer protection legislation, education programs, prison reform, the Trans-Alaska pipeline, airport noise, disease control, infant nutrition, recreation facilities, labor and manpower training programs, and housing programs.(4)

The common elements of cost-benefit analysis are applicable to all areas. There are four main stages: identification, classification, quantification, and presentation. Each of these stages presents its unique problems to the analyst, especially since the work of various participants and disciplines in a project must be combined.(5) In the health, safety and environmental area, quantification of health and human welfare benefits has proved to be an especially controversial topic.(6) Nevertheless, it is worthwhile to consider the application of cost-benefit analysis to regulation in that area in order to improve the quality of regulatory decisions, and to introduce discipline and rigor in the making of those decisions.

In authorizing federal agencies to regulate exposure to hazardous substances in the health, safety and environmental area, the Congress has been far from consistent in providing economic guidelines for such regulation. Thus, some statutes permit regulatory agencies to consider the economic consequences of the regulations: these include the Toxic Substances Control Act of 1976, the Consumer Products Safety Act of 1972, the Resource Conservation and Recovery Act of 1976, and the Comprehensive Environmental Response, Compensation and Liability Act of 1980. In some statutes, however, the ability of the agency to consider the economic effects of rules or to balance the costs against benefits is less clear; examples of such statutes are the Occupational Safety and Health Act of 1970, and the Clean Air Act amendments of 1970 and 1977.(7)

Though the statutory direction for including cost-benefit analysis in the regulatory process is far from consistent, Presidents since Gerald Ford have endeavored by other means to compel its use. In 1974, President Ford by Executive Order required that "promulgation of rules by executive agency must be accompanied by a statement which certifies that the inflationary impact of the proposal has been evaluated."(8) In 1978, his successor, President Jimmy Carter, issued a subsequent Executive Order setting forth as a requirement of significant regulations that each issuing agency perform a regulatory analysis thereon.(9) That analysis should include not only a description of the major alternative ways of dealing with the problem but also an economic analysis of each of these alternatives and a detailed explanation of the reasons for choosing one alternative over the others.

It is thus evident that the President as well as Congress and the agencies it has authorized to regulate have begun to recognize the need to grapple with the societal costs that health, safety and environmental regulations incur in achieving their beneficial objectives.

The Societal Background for Comparing Regulatory Costs and Benefits Before going on to discuss the new rules for evaluating regulations, it is worthwhile to consider the social atmosphere in which the cost-benefit debate is being conducted.

For the last several years, there has been increasing, though minority, public concern over the growth of regulation of society. As Table I reveals, over the last five years the public has become increasingly critical of regulation.

Table I. Perception of Problems with "Big Government" (10) (General Public)

We hear a lot of talk these days about the problem of 'big government.' Which of the following problems do you think of when you think about what's wrong with our government today?

Perceived Problems	1976	1978	1980	1981
Red tape & too much paperwork	57%	54%	63%	67%
Too much regulation of citizen's private lives	25%	24%	34%	44%
Invasions of privacy	33%	34%	41%	42%
Too much regulation of free market system	21%	20%	28%	35%

Thus, currently about a third of the public believes that the free market system is regulated too much, up from about a fifth five years ago.

Moreover, there is general public recognition that while the costs of government regulation increase business costs, those costs are passed on to the consumer. Table II reveals that this perception has been relatively unchanged over the last six years.

Table II. Public Understanding of the Costs of Regulation (11) (General Public)

In your opinion, does conforming to government standards for clean air, greater product safety, etc., involve extra spending for business?

	1975	1977	1979	1981
Yes	83%	85%	71%	78%
No	8%	9%	16%	14%
Don't know/no answer	9%	6%	13%	8%
(Among those answering "yes" above) Do you feel that business:				
Reduces its earnings in order to get the money to conform to these standards	5%	6%	5%	9%
Passes the costs on consumers in the form of higher prices	80%	80%	81%	82%
Don't know/no answer	15%	14%	14%	9%

While around three quarters of the public recognize that government regulation increases business costs, more than three quarters of these or about two thirds of the public understand that these costs are reflected in higher prices.

In light of these data, a fundamental issue is therefore whether the public believes regulatory benefits outweigh the problems created by regulation. The data show that, as Table III indicates, a majority see health, safety and environmental regulations as being generally beneficial.

Table III. Benefits Versus the Problems of Regulations(12)
(General Public 1981)

Here is a list of areas where regulations have been passed in the last decade or two. Tell me for each one whether you think, on balance, the benefits have outweighed the problems, or the problems have outweighed the benefits.

	Benefits Outweigh the Problems %	Problems Outweigh the Benefits %	Don't Know %
Food safety	74	1	7
Product safety	68	23	9
Worker safety	66	24	10
Water pollution	60	33	7
Air pollution	56	37	8
Industrial waste disposal	53	35	12

Conviction that food safety regulatory benefits outweigh the problems is, however, far greater than that for industrial waste disposal.

Finally, there seems fairly reliable evidence that the American public believes that the government should consider regulatory costs when issuing new rules. In a recent survey, Cambridge Reports, Inc., found that more than three times as many Americans believed that cost should be considered as thought it should be ignored, as Table IV indicates.

Table IV. Consider Costs of Regulation?(13)
(General Public)

Some people say that the government should consider how much a new regulation will cost consumers before they decide to make it a law. Other people say that when it comes to protecting consumers, the government should not even consider how much it might cost. Which of these views is closer to your opinion?

	1979	1981
Government should consider cost	70%	71%
Government should not consider cost	18%	19%
Don't know	13%	10%

Thus, it appears that there is public support for agency consideration of regulatory costs in the issuance of regulations.

From these data, some conclusions can be drawn. The first

is that concerns over societal controls appear to be increasing. Second, though the public understands that consumers ultimately bear regulatory costs, for health, safety and environmental controls they believe that regulatory benefits currently outweigh regulatory problems. Nevertheless, a strong public majority believes that regulatory costs should be considered in issuing regulations.

<u>The 1981 Executive Order</u> Shortly after taking office, President Reagan established the Task Force on Regulatory Relief under the chairmanship of Vice President Bush. On February 17, 1981 the President authorized broad regulatory oversight for the Task Force, working with the Office of Management and Budget. The document providing that authorization is Executive Order 12291, whose stated purposes are to reduce the burdens of existing and future regulations, increase federal agency accountability for regulatory actions, minimize duplication and conflict of regulations, and insure well-reasoned regulations.(14)

Basically, the Executive Order requires that in issuing new regulations and in reviewing old ones, the issuing agency undertake regulatory action only when

1. A need for regulation is adequately demonstrated;
2. The potential benefits outweigh the potential costs and adverse effects; and
3. The most cost-effective and least burdensome approach is established.

Thus, the Executive Order requires that agencies affected by the order employ cost-benefit criteria in developing and issuing regulations. The tool to be applied for that purpose is the Regulatory Impact Analysis.

The nature of the Regulatory Impact Analysis (RIA) is specified in Section 3 of the Executive Order. In general, such analyses are required only for rules which the issuing agency determines are major rules. The order defines a major rule as any regulation likely to result in

1. An annual effect on the economy of $100 million or more;
2. A major increase in costs or prices for consumers, individual industries, federal, state, or local government agencies, or geographic regions, or
3. Significant adverse effects on competition, employment, investment, productivity, innovation, or on the ability of United States-based enterprises to compete with foreign-based enterprises in domestic or export markets.

Thus, the process of determining whether or not a proposed rule is a major rule requiring an RIA will develop considerable data required for the cost side of any subsequent cost-benefit analysis.

Information required in the RIA includes

1. A description of the potential benefits of the rule, including any beneficial effects that cannot be quantified in monetary terms, and the identification of those likely to receive the benefits;

2. A description of the potential costs of the rule, including any adverse effects that cannot be quantified in monetary terms, and the identification of those likely to bear the costs;
3. A determination of the potential net benefits of the rule, including an evaluation of effects that cannot be quantified in monetary terms;
4. A description of alternative approaches that could substantially achieve the same regulatory goal at lower cost, together with an analysis of this potential benefit and costs and a brief explanation of the legal reasons why such alternatives, if proposed, could not be adopted; and
5. Unless covered by the description required under (preceding) paragraph (4) an explanation of any legal reasons why the rule cannot be based on the requirements set forth in Section 2 of this Order.

The RIA thus encompasses both the information required for a cost-benefit analysis of the proposed rule, and for a determination of the cost-effectiveness of the regulatory approach incorporated therein.

It is important to recognize that the required descriptions of both costs and benefits include both effects quantifiable in monetary terms, effects quantifiable in other than monetary terms, and effects that cannot be quantified. In the health, safety and environment area, these unquantifiable effects will include such difficultly-addressable elements as aesthetic values; disease, pain and suffering and their alleviation; and, ultimately, the value of life. It has, of course, been pointed out that there are important differences between economic regulation and regulation dealing with health, safety and environment. As one writer has pointed out, the benefits side of cost-benefit studies in this area includes improved quality of life as well as positive economic side effects, and the former defy accurate estimation. Moreover, the comparison of costs and benefits is beset by serious methodological difficulties and requires the analyst to make value-laden assumptions.(15) Accordingly, the application of Regulatory Impact Analysis to health, safety and environmental regulation will not only present methodological challenges but will no doubt generate extensive controversy as well.

Regulatory Impact Analyses so prepared are required to be reviewed by the Director of the Office of Management and Budget, subject to the direction of the Presidential Task Force on Regulatory Relief. Thus, Executive Order 12291 affords two regulatory review levels, one by OMB and potentially one by the Task Force. The OMB, as an agency of the Executive Office of the President, is in a position to ensure that both the letter and the spirit of Executive Order 12291 are followed in RIAs produced by the several agencies.

Executive Order 12291 is not, however, generally applicable to all federal agencies. In fact, it specifically exempts those federal agencies designated by statute as "independent regulatory agencies." Independent regulatory agencies thus not included in the requirement to conduct RIAs and which have health, safety and environmental responsibility include the Mine Enforcement Safety and Health Review Commission, the Nuclear Regulatory Commission, and the Occupational Safety and Health Review Commission. Moreover, in their 1981 Cotton Dust decision, the U.S. Supreme Court has determined that the OSHA statute does not require cost-benefit determinations when regulations are issued thereunder.(16)

Just what information is required in RIA, and what techniques are required to produce those data for each affected agency remain unclear. At the time of writing this paper, it was generally understood that EPA and other affected agencies were preparing guidelines for preparation of RIAs that would afford additional insight into their approach to these documents.

The proposal to employ cost-benefit methods in regulatory impact analysis is not without its critics. In a November, 1981 address at a University of Virginia Law School meeting, U.S. Senator Robert T. Stafford opposed the use of cost-benefit analysis in environmental issues. His comments about the cost-benefit analysis are summed up in the following points:

- In his opinion, monetizing costs and benefits converts an intangible right-health to a property right, which is then "involuntarily alienated."
- Under such a system, the government would create a system in which the polluter would have the right to injure others because it would cost him too much to avoid harming them.
- The system would be skewed in favor of pollution because benefits are difficult to determine but costs are easily calculated.

He further suggests that the use of economic analysis to evaluate health-related rules jeopardized all other rights in American society.(17)

Since this view is shared by other commentators and constituencies, it is likely that the debate over application of Executive Order 12291 will be a long one.

The Chemical Industry Position As this paper has shown, cost-benefit analysis has now been incorporated into the regulatory process of many, though not all, agencies dealing with health, safety and the environment. The American chemical industry is, of course, profoundly affected by those agencies, since their rules deal with its operations, its products, its wastes and its relation to the communities in which chemical plants are located. It is therefore worthwhile to examine what the position of the industry is toward cost-benefit analysis in regulatory impact determinations.

The American chemical industry includes, of course, a wide range of companies in a variety of activities and currently employs over a million workers, about 1% of the U.S. labor force. A prominent voice for that industry is the Chemical Manufacturers Association, a trade association including around two hundred member companies. As part of its activities, CMA has taken an increasingly strong advocacy role, speaking out responsibly on national issues affecting the chemical industry in general.

In late 1980, a special CMA work group concluded that cost-benefit analyses were being increasingly employed in regulatory decision-making. Such analysis were being performed in inconsistent ways, and were flawed by a lack of scientifically-acceptable data and inadequate definition of the terms employed. Moreover, there seemed to be no agreed-upon methodology for the conduct of these analyses. At that time, CMA established a committee of industry experts who embarked upon a concentrated effort to develop principles and definitions for use in cost-benefit studies in the health, safety and environment field. As a result of this effort, during 1981, the Association developed a policy for regulatory impact analysis of health, safety and environmental regulations. That policy is now being made public.(18)

In that policy, CMA has unequivocally stated that regulatory agencies should perform regulatory impact analyses to make governmental decision-making processes more effective. The Association believes that improved analysis at the beginning of a regulatory proposal will allow workable and effective rules to be in place sooner. Economic, scientific and technical issues should be included in the analysis.

The Chemical Manufacturers Association has recommended the following guidelines for conducting RIAs:

- Regulations should be adopted when (1) a need for regulation has been demonstrated, (2) costs bear a reasonable relationship to benefits, and (3) the most cost effective approach is adopted;
- Regulatory impact analysis should not include quantification of intangibles in monetary terms;
- Regulatory agencies should use "good science" in defining both the need for a regulation and the benefits in terms of risk reduction it will provide; and
- Regulatory agencies should evaluate alternative approaches to regulation.

These guidelines thus introduce some new concepts into the RIA procedure, as well as support concepts already embodied in Executive Order 12291.

CMA believes that regulation in the health, safety and environmental area should be adopted only when it materially reduces real hazards. Regulations should be adopted only where they will significantly reduce risk. Moreover, regulations should not be used to induce small changes or to reduce already minor risks. Justification for a regulation should be based on scientific data that clearly identify the hazard to be reduced and show to what extent the regulation will reduce the hazard.

Calculation of costs and benefits should go beyond mere accounting procedures. CMA believes that the anticipated cost of a regulation should include both the direct costs of complying with it and its indirect costs throughout the economy. Similarly, the benefits to be included are the direct and indirect benefits of the regulation.

CMA also supports the cost-effectiveness requirement of Executive Order 12291. It believes that regulatory agencies should analyze the potential costs and benefits of reasonable alternatives for achieving regulatory goals. Non-regulatory approaches, such as economic incentives, can be more effective and less costly than regulations.

Agencies should also consider alternative methods of regulatory control. Such alternatives might include flexible compliance deadlines, performance standards, variances and exceptions.

Finally, CMA urges that a regulation should take the least burdensome approach that will achieve its goals. Resources are wasted whenever a regulation imposes requirements not directly related to its objectives.

Two points not addressed explicitly in the Executive Order that are dealt with by CMA are the value of human life, and the importance of good scientific information. CMA has stated in its position that regulatory agencies should not place a dollar value on human life, other health effects such as pain and suffering, or aesthetics. Such quantification is not meaningful to society, and the use of mechanistic cost/benefit ratio for decision-making would be unwise. Regulatory impact analysis should provide decision-makers with as much information as practicable to ensure that regulations express human as well as economic values.

Like many other commentators, CMA has been concerned over the poor quality of data that have been employed for making rules in the health, safety and environmental areas. Accordingly, CMA urges in its guidelines that agencies should use quantitative risk assessments that are based on substantial evidence. Unsupported assumptions or seriously flawed scientific studies form a poor basis for regulation.

Analyses should be reviewed by independent scientists to ensure that the data are valid and that interpretations are correct. Thus, independent peer review is an important element of the CMA position, though one not covered by the Executive Order.

Issuance of the policy position of the Chemical Manufacturers Association is only an early step in the role CMA expects to play in the national debate over cost-benefit analysis in regulatory affairs. A number of CMA committees are actively engaged in examining the many issues involved, and worthwhile contributions to the debate from the chemical industry are likely.

Conclusions

The issuance of Executive Order 12291 has now made cost-benefit analysis a necessary part of the issuance of some federal rules. At this time, bills in both houses of Congress are being considered to expand the scope of cost-benefit analysis and to elevate it to legislative status.

Central issues in the debate include that of how to deal with such values as human pain and suffering and ultimately with human life. As yet, no agreed-upon methodology for carrying out the necessary calculations exists. Moreover, the field of debate has to date been occupied mainly by scientists, engineers and economists. Accordingly, it can be expected that, as such new participants as the moral philosophers and social scientists join the discussion, new issues will be raised and new insights gained.

Some conclusions can nevertheless be drawn. It seems clear from the survey research data that the American public is concerned over the increase in social regulation, and that there is growing interest in introducing the cost factor into agency considerations. It can reasonably be concluded that, so long as Executive Order 12291 requires regulatory impact analysis, cost-benefit analysis will play that function. Nevertheless, the nature of the debate is very likely to change in ways not yet anticipated by the present participants.

Literature Cited

1. See in this regard, "The Point Is...A Summary of Public Issues Important to the Dow Chemical Company," No. 48, January 11, 1982 (Dow Chemical Co., Midland, Mich.).
2. See for example AAAS Special Symposium No. 65, Risk in the Technological Society, Ed. by C. Hohenemser and J.X. Kasperson, (Westview Press, Boulder, Colo. 1982).
3. Peter G. Sassone and William A. Schaeffer, Cost-Benefit Analysis: A Handbook, p.3 (Academic Press, New York, 1978).
4. Lee G. Anderson and Russell F. Settle, Benefit-Cost Analysis: A Practical Guide, p.1 (Lexington Books, D.C. Heath & Co., Lexington, Mass., 1977).
5. Anderson and Settle, loc. cit., pp.1-2.
6. See for example Steven Kelman, "Cost-Benefit Analysis, An Ethical Critique," Across the Board, pp.74-82, July-August 1981.
7. The writer is indebted to Mr. T. H. Rhodes who conducted this analysis. And cf. Michael S. Baram, "Cost-Benefit Analysis: An Inadequate Basis for Health, Safety and Environmental Decision-Making" Ecology Law Quarterly 8 pp.473-531 (1980), Footnote 1.

8. Executive Order No. 11821, Nov. 29, 1974, 39 F. R. 41502, Amended by Executive Order No. 11,949, Dec. 31, 1976, 42 F. R. 1017.
9. Executive Order No. 12044, March 23, 1978, 43 F. R. 12661.
10. Yankelovich, Skelly & White, Inc. Corporate Priorities 1981, Table A-1.6 (New York, 1981).
11. Yankelovich, Skelly & White, Inc., loc. cit., Table A-2.5.
12. Yankelovich, Skelly & White, Inc., loc. cit., Table A-2.16.
13 Cambridge Reports, Inc., The Cambridge Report 27, p.214 (Cambridge Mass., 2d Quarter 1981).
14. 46 Federal Register No.33, Feb. 19, pp.13193-13198.
15. Nicholas A. Ashford, "The Limits of Cost-Benefit Analysis in Regulatory Decisions," Technology Review pp.70-72, May 1980. And cf. M. S. Baram, loc. cit.
16. AMERICAN TEXTILE MANUFACTURERS INSTITUTE, INC., ET AL. v. DONOVAN, SECRETARY OF LABOR, ET AL., No. 79-1429., SUPREME COURT OF THE UNITED STATES, 49 U.S.L.W. 4720 June 17, 1982, together with No. 79-1583, National Cotton Council of America v. Donovan, Secretary of Labor, et al., also on certiorari to the same court, 101 S. Ct. 2478, 69 L. Ed. 2d. 185.
17. Remarks of U.S. Senator Robert T. Stafford at the Conference for Toxic Substances Pollution, The Environmental Law Institute, University of Virginia Law School, Nov. 23, 1981.
18. Chemical Manufacturers Association, Policy for Regulatory Impact Analysis of Health, Safety and Environmental Regulations, (Washington, D.C. 1981).

RECEIVED October 14, 1983

INDEXES

Author Index

Subject Index

F

G

H

I

J

K

L

Production and indexing by Anne Riesberg
Jacket design by Anne G. Bigler

Elements typeset by Hot Type Ltd., Washington, DC
Printed and bound by Maple Press Co., York, PA